Dr. Math® Gets You Ready for

ALGEBRA

Coming soon . . .

Dr. Math® Explains Algebra

• • • • • • • • • •

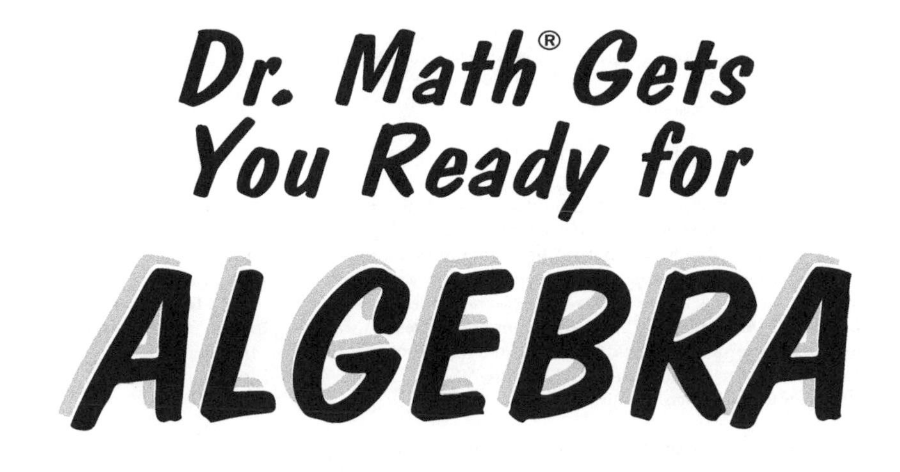

Learning Pre-Algebra Is Easy! Just Ask Dr. Math!

THE MATH FORUM

Cartoons by Jessica Wolk-Stanley

Published by Jossey-Bass
A Wiley Imprint
989 Market Street, San Francisco, CA 94103-1741 www.josseybass.com

Published simultaneously in Canada

Design and production by Navta Associates, Inc.

Library of Congress Cataloging-in-Publication Data

Dr. Math gets you ready for algebra : learning pre-algebra is easy! just ask
Dr. Math! / the Math Forum@Drexel
p. cm.
Includes index.
ISBN 978-0-471-22556-0 (pbk.)
1. Algebra. I. Title: Dr. Math gets you ready for algebra. II. Math Forum@Drexel.

QA152.3.D7 2003
512—dc21 2003043278

FIRST EDITION
PB Printing 10 9

Contents

Acknowledgments

Suzanne Alejandre and Melissa Running created this book based on the work of the Math Doctors, with lots of help from Math Forum employees, past and present:

Ian Underwood, Attending Physician

Sarah Seastone, Editor and Archivist

Tom Epp, Archivist

Lynne Steuerle and Frank Wattenberg, Contributors to the original plans

Jay Scott, Math Consultant

Kristina Lasher, Associate Director of Programs

Stephen Weimar, Director of the Math Forum

We are indebted to Jerry Lyons for his valuable advice and encouragement. Our editors at Wiley, Kate Bradford and Kimberly Monroe-Hill, have been of great assistance.

Our heartfelt thanks go out to the hundreds of Math Doctors who've given so generously of their time and talents over the years, and without whom no one could Ask Dr. Math. We'd especially like to thank those doctors whose work is the basis of this book:

Anthony Hugh Back, Pat Ballew, Guy Brandenburg, Daniel Brown, Roger Cappallo, Jaime Carvalho e Silva, Gabe Cavallari, Michael F. Collins, Bob Davies, Tom Davis, Sonya Del Tredici, Tim Erickson, James Ezick, C. Kenneth Fan, Barney Feist, Derrel Fincher, Sydney Foster, Sarah Seastone Fought, James Gill, Darren Glass, Margaret Glendis, Brian Gordon, Byron Holz, Gerald Kulm, Elizabeth Li, Alan Lipp, Ethan Magness, Douglas Mar, Jerry Mathews, Micah McDaniel, Naomi Michlin, Josh Mitteldorf, Paul Narula, Elise

Fought Oppenheimer, David Peterson, Richard Peterson, Michael R. Pipe, Andree Reno, Jodi Schneider, Keith Evan Schubert, Gary Simon, Allan Simonsen, Mark Snyder, Rachel Sullivan, Ian Underwood, Joe Wallace, Robert L. Ward, Stephen Weimar, Larry White, John Wilkinson, and Ken Williams.

Drexel University graciously hosts and supports The Math Forum, reflecting Drexel's role as a leader in the application of technology to undergraduate and graduate education.

Dr. Math® Gets You Ready for ALGEBRA

Introduction

Taking your first algebra course is a big step in your math learning experience. You jump from the concrete world of numbers and real objects you recognize to the abstract world of letters and symbols. To quote Dr. Math®, "Algebra is the class where you learn how to work with unknown quantities." You may be wondering what math you'll need to know to succeed in this new subject. The reassuring news is that almost every student faces this same situation and feels just like you. You are definitely not alone. For years students have been asking Dr. Math questions about how to figure out math problems using algebra, and Dr. Math has been helping them by replying with clear explanations and helpful hints. When you have spent time working through this book and perhaps checking out some of the Web sites we point to, you will be prepared to make the big leap into your algebra course.

Dr. Math® Gets You Ready for Algebra is organized into five parts covering topics in a sequence that follows the outline of the beginning of a standard first-level algebra course. The letters that you will read were written by actual students who were having difficulty understanding the basic math concepts that are used as the basis for algebra. The answers, written by Math Doctors, are from the Math Forum's Ask Dr. Math service. Math Doctors are trained volunteers drawn from a pool of college students, mathematicians, teachers, and professionals from the mathematical community.

We start by looking at general ways to think about and work with numbers. Algebra gives us a way to work with numbers whose values we don't yet know. We represent these values with variables. Just as you learned how to work with simple numbers in elementary school, you'll learn how to do the same sorts of operations with variables—addition, subtraction, multiplication, and so on. The first part also covers exponents, large and small numbers, and how variables work with them all.

After considering some of the fundamental operations, we move on to learning about integers in Part 2. Students starting an algebra class often get confused in working with positive and negative numbers. In the explanations on integers, you will find connections to the real world and some reasons why negative numbers are important. We give you some tips and tricks to help you make sense of them throughout your work.

Part 3 will further expand your thinking to include real numbers. In addition to counting or natural numbers, whole numbers, and integers, we'll also be thinking about rational and irrational numbers, and some useful techniques for working with them. A special irrational number that you have probably heard about is *pi*. Why is it helpful to learn about pi and other irrational numbers? As you will see in Part 3, all of these numbers have essential connections to the world around us.

Part 4 is about the meat of algebra—that is, working with equations using variables. We also talk about how to plot equations on a graph, which will be useful not only in algebra but also in geometry. We discuss what it means for equations to be equivalent and some techniques for solving equivalent equations.

Finally, Part 5 gives you a glimpse of bigger things to come as you work with word problems and real-life situations. You'll learn how to set up equations and "translate" them from English to algebra and back again. The tools you'll learn here will help you start your journey through your first algebra course.

Dr. Math welcomes you to the world and language of algebra!

PART 1

Fundamental Operations

Operations are the arithmetic skills introduced and practiced in elementary school. The *fundamental* operations are addition, subtraction, multiplication, and division. Exponentiation is also an operation. In algebra, the fundamental operations are as important as they are in arithmetic. In fact, if you ever want to check your algebraic work by substituting a number for the variable, you'll be reminded of the arithmetic exercises that look more familiar.

Clive and Carissa have a lot of questions about what they're learning. In this part, Dr. Math explains

- Introduction to algebraic thinking
- Variables
- Exponents
- Large and small numbers
- Order of operations
- Distributive property and other properties

1 Introduction to Algebraic Thinking

Algebraic thinking is the bridge between arithmetic and algebra. Representing, analyzing, and generalizing a variety of patterns with tables, graphs, words, and, when possible, symbolic rules are all part of thinking algebraically.

What Is Algebra?

```
Dear Dr. Math,
What is algebra?
Yours truly,
Clive
```

Hi, Clive,

Algebra is like arithmetic, but in algebra some of the numbers have names instead of values. For example, if I ask you something like

$$3 + 4 \cdot 5 - 6 \div 3 = ?$$

you can just apply the operations in the correct order to get

$$3 + 4 \cdot 5 - 6 \div 3 = ?$$
$$3 + 20 - 6 \div 3 = ?$$

$$3 + 20 - 2 = ?$$
$$23 - 2 = ?$$
$$21 = ?$$

Now, suppose that instead I ask you something like

$$(x + 3) \cdot (x + 4) = 42$$

You can't apply the operations because you don't know the values of the numbers. What's $x + 3$? It depends on the value of x, doesn't it?

In this case, you might start guessing possible values for x that would make the equation true:

$x = 1?$ $(1 + 3) \cdot (1 + 4) = 4 \cdot 5 = 20$ (No.)
$x = 2?$ $(2 + 3) \cdot (2 + 4) = 5 \cdot 6 = 30$ (No.)
$x = 3?$ $(3 + 3) \cdot (3 + 4) = 6 \cdot 7 = 42$ (Yes!)

However, suppose the problem changes to

$$(x + 3) \cdot (x + 4) = 35.75$$

Now it becomes a lot harder to guess an answer. Algebra gives you a set of tools for figuring out the answers to problems like this without having to guess.

This becomes more and more important as you start using more complicated equations involving more than one variable.

— ***Dr. Math, The Math Forum***

Rx In case you've forgotten, here's the correct **order of operations** for any equation:

1. Parentheses or brackets
2. Exponents
3. Multiplication and division (left to right)
4. Addition and subtraction (left to right)

For more about this, see section 5 in this part.

What Is Algebraic Thinking?

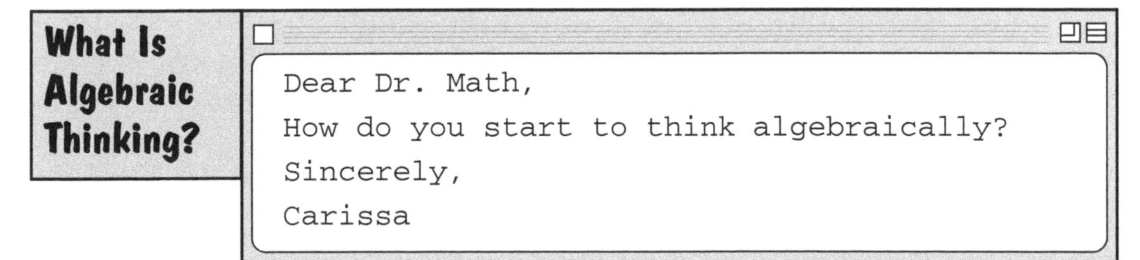

Dear Dr. Math,
How do you start to think algebraically?
Sincerely,
Carissa

Hi, Carissa,

Well, you already know about multiplication, division, addition, and subtraction. One day (a long, long time ago) somebody—let's call him or her Pat—who knew all of those things was sitting around thinking about addition.

Pat knew that $3 + 4 = 7$.

Then Pat asked, "What would happen to the equation if I added 1 to both sides?" Pat tried it and got

$$3 + 4 + 1 = 7 + 1$$

Pat realized right away that this new equation was also true. Then Pat went back to the original equation of $3 + 4 = 7$, decided to subtract 3 from both sides, and got

$$3 + 4 - 3 = 7 - 3$$

Pat then did some arithmetic and ended up with $4 = 4$.

Right away, Pat realized that this technique could be applied to different types of equations. Pat asked, "What if I didn't know one of the numbers?"

Pat was already familiar with equations like $3 + 4 = ?$ and knew that you could **solve** those equations.

Pat decided to try something a little different: $? - 4 = 7$. Pat knew from before that you can add or subtract the same number from both sides of an equation (see above) and still have a true equation. So Pat added 4 to both sides of this equation and got

$$? - 4 + 4 = 7 + 4$$

After a little bit more arithmetic, Pat ended up with $? = 11$. If you keep thinking like this, and instead of using ? you use x or y or a to stand for the missing number, that means that you are starting to think algebraically.

— Dr. Math, The Math Forum

Variables

A **variable** is a symbol like x or *a* that stands for an unknown quantity in a mathematical expression or equation. If you remember that the word *variable* means *changeable,* then it is a little easier to remember that the value of the x or *a* changes depending on the situation.

For example, what if you are thinking about the number of tires you need for a certain number of cars? You know that 4 tires are needed for each car. You can write $4c = t$, where c is the number of cars, *t* is the number of tires, and 4c means 4 times c. If there are 25 cars, you can figure out that $4(25) = 100$, so you will need 100 tires. If there are 117 cars, you know that $4(117) = 468$ and you will need 468 tires. Because the number of cars can change but the relationship between the cars and tires stays the same, the formula $4c = t$ is a useful way to explain the general situation.

An algebraic **expression** is like a *word* or *phrase,* and an **equation** is like a *sentence.* For example, $12c$ might represent a quantity of eggs. If c is the number of cartons of eggs and there are 12 eggs in each carton, you can see that $12c$ is a useful way of *expressing* the total quantity of eggs.

If you have another number or expression, you can relate it to your first expression in an equation. For example, let $12c = 60$. This is an algebraic equation with one variable. Solving this kind of equation reveals $c = 5$. Thus we have 5 cartons.

Whereas equations can be *solved,* expressions can only be *evaluated* or *rearranged* or *simplified.* Note also the distinction between having $12c$ by itself (an expression) and relating it to another expression using an equal sign: $12c = 60$ (an equation).

By the way, in an expression like $4c - 3$, 3 is called a **constant,** because it doesn't vary. The 4 changes along with the variable it multiplies and is called the coefficient of c.

What Are Variables For?

Dear Dr. Math,

Why is it important to be able to figure out the values of variables? We've been doing that in our math class for more than half a year and I was just wondering why we are doing it.

—Carissa

Hi, Carissa,

This is a very perceptive question.

Variables are important for a couple of reasons, which we might call *planning* and *analysis.*

Think about planning a dinner party. Let's say you know that you'll need one-half of a chicken for each adult and one-quarter of a chicken for each child; you'll need one bottle of wine for every three adults and one bottle of soda for every five children; you'll need a half pound of potatoes for each chicken that you have to cook; you'll need one pie for six adults and one bowl of ice cream for each child; and so on.

But you don't yet know how many people you're going to invite. Variables let you set up a description of the situation (i.e., an equation) such that you can plug in two numbers (the number of adults and the number of children) and get back other numbers that you'll find useful: how many chickens to buy, what the total cost will be, and so on. If you decide at the last minute that you want to add three more guests, you don't have to start your calculations from scratch—you just change the values coming in and the equations will tell you how to change the values at the other end.

This, by the way, is why they are called *variables*—they tell you how some quantities vary in response to changes in other quantities.

Note that a dinner party isn't all that complicated, so it's almost not worth the effort of setting up equations to solve the problems. But when you get to something more complicated—like trying to plan the flight of an airplane or run an entire airline—it becomes absolutely necessary to use variables. A big part of running any business is being able to figure out your potential costs in any situation, because that tells you how much you need to charge for goods and services in order to make enough money to stay in business.

So, that's planning. What about analysis? Well, analysis is just planning in reverse. If you know how many people to invite, you can figure out how much money you'll have to spend. That's planning. If you know how much money you spent, you can figure out how many people you invited. That's analysis. The beauty of variables is that in most cases you can use the same equations to go in either direction—to predict what's going to happen or to understand what already happened.

The planning aspect tends to be more useful in things like business or construction or engineering, where you have to decide what's going to happen. The analysis aspect tends to be more useful in science, where you don't get to decide what happens (the world behaves the way it behaves, whether you like it or not) but would like to understand it anyway, whether due to curiosity or because you'd like to use that understanding to make your planning more accurate.

—Dr. Math, The Math Forum

Using Variables

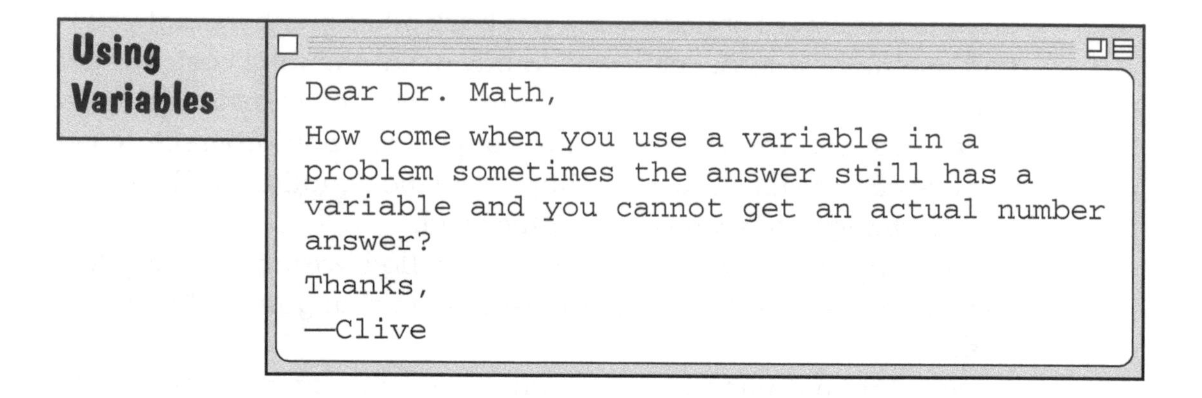

Dear Dr. Math,

How come when you use a variable in a problem sometimes the answer still has a variable and you cannot get an actual number answer?

Thanks,

—Clive

Dear Clive,

I assume you are not talking about making a mistake in solving the problem. If there is only one variable in the original equation, then either you can solve it with a numerical answer or you simply can't solve it—there would be no actual solution that still involved the variable.

But if you are given an equation with two variables in it, like $w = \frac{24}{h}$, and are told to solve it for one of the variables, say $h = \frac{24}{w}$, then the other variable will still be there. In this case, you are simply rearranging a formula for a different use. As given, the formula lets you get the width of a rectangle given its height. After you solve for h, it lets you find the height of a rectangle given its width. You don't know either one yet, but if I gave you a height, you could plug it right into this formula. If you hadn't already solved for h, you would have to put my value into the original equation for w and then solve that for h.

So, there are two ways a variable can be used. Sometimes it is an unknown, which you want to figure out from the equation. Other times it just stands for a value that you don't know now but will know later, like w in my example. Then you just work with it as if it were a value but without being able to do the calculations. When you're done, you can replace it with any value.

—***Dr. Math, The Math Forum***

Writing Expressions with Variables

Dear Dr. Math,

Here's a problem that I'm having trouble with:

> Write an expression that represents a \$500 donation plus \$5 for every event. Let n represent the number of events.

I do not understand what the problem wants me to write.

Yours truly,

Clive

Hi, Clive,

An expression is a collection of numbers and variables connected by arithmetic operations (add, multiply, etc.), so if you worked out all the arithmetic (which is called **evaluating** the expression), you would get a number. In this case, the number would be the total payment.

Let's say you knew there were 4 events. Could you then work out how much to pay? It would be

$$500 + 5 \cdot 4$$

in dollars. This is an expression. You can work it out and get the answer 520.

In fact, you don't know how many events there are. But whatever that number turned out to be, you could put it in place of the 4 in the expression and work it out in just the same way. So we use a letter as a name to stand for whatever number we will end up putting there. This is a variable. It is sort of a placeholder for a number.

You were asked to use the letter n to represent the number of events. That means n will be our variable and we can put it in place of the 4, like this:

$$500 + 5 \cdot n$$

When multiplying by a variable, you don't need to write the "·". You can just write

$$500 + 5n$$

I hope this helps you work out other problems in writing expressions.

—Dr. Math, The Math Forum

Understanding Variables

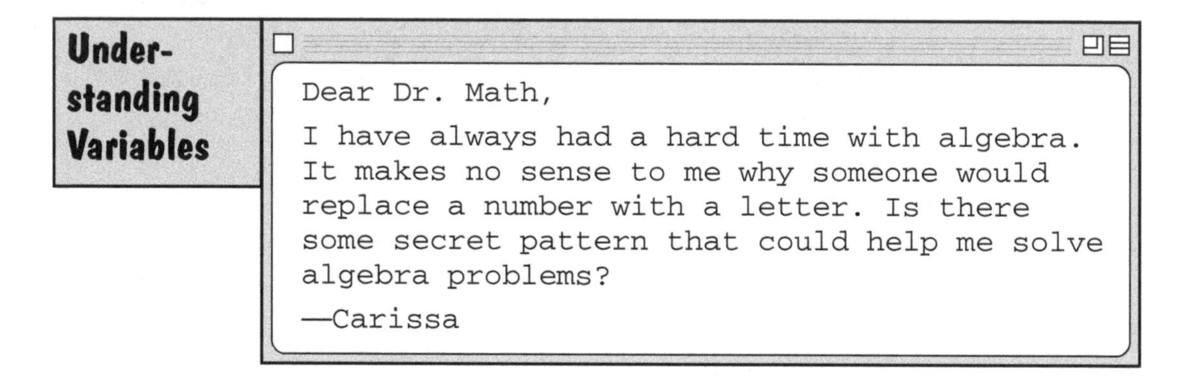

Dear Dr. Math,

I have always had a hard time with algebra. It makes no sense to me why someone would replace a number with a letter. Is there some secret pattern that could help me solve algebra problems?

—Carissa

Hi, Carissa,

You're doing something that's a little like algebra whenever you use a pronoun. You could have written this:

> Carissa has always had a hard time with algebra. It makes no sense to Carissa why someone would replace a number with a letter. Is there some secret pattern that could help Carissa solve algebra problems?

When you wrote to us, you used the pronouns *I* and *me* to take the place of your name. A variable is like a pronoun: it's a way of talking about a number without calling it by name.

We don't generally replace a number with a letter; more often we don't know the number yet, so we just give it a nickname (like x) and work with it until we can replace the letter with the right number.

The great discovery that made algebra possible was the realization that even if you don't know what a number is, you can still talk about it and know certain things about its behavior; for example, no matter what the number is, if you add 2 to it and then subtract 2 from the result, you'll have the same number you started with. We can say

$$x + 2 - 2 = x \text{ for any } x$$

Here's one secret that may help you: when you see an equation that confuses you, try putting an actual number in place of the variable and see if it makes sense. For example, in what I just wrote, you could try replacing x with 47:

$$47 + 2 - 2 = 47$$

It works! Now think about why it works: 47 plus 2 means you've gone 2 units to the right; minus 2 takes you 2 units back to where you started. It doesn't matter that the place at which you started was 47; adding 2 and subtracting 2 undo one another.

I'll take you one step deeper into algebra and actually solve an equation. Let's say we're told that

$$3x - 2 = 7$$

In words, I can say, "I have a secret number. If I multiply it by 3 and then subtract 2, I get 7. What is it?" (Notice how I used pronouns to stand for the number.) In order to solve this, I can think of it as if the x were a present someone wrapped up for me. First, someone put on some "times 3" paper and then over that some "–2" paper. The package I was given is a 7. I want to unwrap it and see what the x is that's inside.

To take off the "–2," I can add 2 (remember what we said before

I think I'm getting the hang of variables now. Here's a practice problem for you, Carissa. Do you know what n is if 3n+5 = 14?

Hmm... I subtract 5 from both sides of the equation and I have 3n = 9. Then I divide both sides of the equation by 3 and that tells me that n = 3. Does my answer check?

Well, let's see: if we go back to the original equation and replace the "n" with a 3, we have **3(3)+5 = 14.** Doing the arithmetic, we can see both sides of the equation are 14.

It checks!

about adding and subtracting 2 to both sides of the equation). It works like this:

$$3x - 2 = 7$$
$$3x - 2 + 2 = 7 + 2$$
$$3x = 9$$

So I've taken off the "–2" paper and what I found inside was a 9. Now we can take off the "times 3" by dividing both sides of the equation by 3:

$$3x \div 3 = 9 \div 3$$
$$x = 3$$

Now the present is unwrapped and we can see what it is. We were able to do all this because we knew how to handle a number without knowing what it was. Of course, since we can always make mistakes, we should check that we're right; let's wrap it back up and see if it's a 7:

$$3(3) - 2 = 9 - 2 = 7$$

Yup! That's what was in the package. And doing this lets us see what was happening to x by putting a real number (the right one) in its place.

—Dr. Math, The Math Forum

3 Exponents

Until about four hundred years ago, nobody used exponents, but they were perfectly able to do mathematics. For example, they would write $5 \cdot 5 \cdot 5 \cdot 5 \cdot 5 \cdot 5$. But between the fourteenth and seventeenth centuries, mathematicians in Europe developed the concept of raised exponential notation. They decided to use exponents to mean how many times they wrote down the number. So, $5 \cdot 5 \cdot 5 \cdot 5 \cdot 5 \cdot 5$ was written quickly as 5^6.

Properties of Exponents

Dear Dr. Math,

I was out of school with the flu and when I came back my class had studied properties of exponents. Will you please help me understand exponents?

Thanks,

Carissa

Hi, Carissa,

I'm going to start from the definition of what an exponent is and show how you can figure out the properties from the definition. So, the first thing to know about **exponents** is that they are just a shorthand notation for a special kind of multiplication. That is, there are times when I need to refer to a number like

$$6 \cdot 6 \cdot 6 \cdot 6 \cdot 6$$

but I don't want to keep writing that down all the time. We have a much nicer notation, which is

$$6 \cdot 6 \cdot 6 \cdot 6 \cdot 6 = 6^5$$

Whenever you see something that looks like

$$a^b$$

you know that it means b copies of a, all multiplied together. Here are some more examples:

$$2^3 = 2 \cdot 2 \cdot 2$$
$$3^2 = 3 \cdot 3$$
$$(a + b)^4 = (a + b) \cdot (a + b) \cdot (a + b) \cdot (a + b)$$
$$(4^5)^2 = (4^5) \cdot (4^5)$$

It's just an abbreviation, like writing *Dr.* instead of *Doctor* or like writing *MA* instead of *Massachusetts.* You need to be very clear on that or nothing else about exponents is going to make sense.

Multiplying Exponents

When multiplying numbers with exponents that have the same base, it's easy to forget whether to add or multiply those exponents. I use examples to help me remember. If I have

$$(a \cdot a) \cdot (a \cdot a \cdot a)$$

that's $(a^2) \cdot (a^3)$. I know the answer has to be a^5 just from counting the a's. Since I can get that 5 by adding the exponents, that tells me

$$a^b \cdot a^c = a^{(b + c)}$$

So, to multiply powers of the same base, we add the exponents.

Here is another property of multiplying exponents. If I have something like

$$(a \cdot a) \cdot (a \cdot a) \cdot (a \cdot a)$$

that's $(a^2)^3$, and I know the answer has to be a^6 (again, just from counting the a's). Since I can get 6 by multiplying the exponents, I know that

$$(a^b)^c = a^{(b \cdot c)}$$

In other words, to raise a power to a power, we multiply the exponents.

Dividing Exponents

If I have an expression like

$$\frac{a \cdot a \cdot a}{a \cdot a}$$

that's the same as a^3/a^2. If I cancel out as many a's as I can from the original problem, I'm left with a single a, which tells me that

$$\frac{a^b}{a^c} = a^{(b - c)}$$

In other words, to divide powers of the same base, we subtract the exponents.

What happens when you flip that fraction upside down? You get

$$\frac{a \cdot a}{a \cdot a \cdot a}$$

which is a^2/a^3. If I cancel out as many a's as I can from the problem, I'm left with $\frac{1}{a}$. From the property above, I know that $a^2/a^3 = a^{2-3} = a^{-1}$, which tells me that

$$a^{-b} = \frac{1}{a^b}$$

Here's another example:

$$\frac{a \cdot a}{a \cdot a} = ?$$

which is the same as a^2/a^2. From the property above, we know that a^2/a^2 should equal $a^{(2-2)}$, which equals a^0. But I know that it has to be 1, since anything divided by itself must be 1. So,

$$a^0 = 1$$

To summarize, here are the interesting properties of exponents. If *a* and *b* are positive integers, then

$$a^b \cdot a^c = a^{(b+c)}$$

$$\frac{a^b}{a^c} = a^{(b-c)}$$

$$(a^b)^c = a^{(b \cdot c)}$$

$$a^{-b} = \frac{1}{a^b}$$

$$a^1 = a$$

$$a^0 = 1$$

$$a^{\frac{1}{n}} = \sqrt[n]{a} = \text{the } n\text{th root of } a$$

Don't forget that these properties apply only if the bases are the same—you can't do anything with $5^4 \cdot 3^2$ except work it out numerically. You can't do anything with $x^3 \cdot y^7$ as it stands.

for any value of a, which seems a little weird. But it follows from the definitions we've been working out, and it doesn't lead to any bizarre consequences, so we just accept it.

You can do a little more with exponents, too. According to the rules that we just figured out, note that

$$a^{\frac{1}{2}} \cdot a^{\frac{1}{2}} = a^{(\frac{1}{2}+\frac{1}{2})} = a^1 = a$$

So, $a^{\frac{1}{2}}$ multiplied by itself is a . . . , which tells me that

$$a^{\frac{1}{2}} = \sqrt{a}$$

$$a^{\frac{1}{3}} = \sqrt[3]{a}$$

and so on.

The main thing you need to do is avoid the feeling that you have to memorize these properties. As you can see, they are all just *consequences* that follow from the definition of an exponent, which is just something that mathematicians invented so that they could be lazy about writing things down. If you really understand the definition, you can make up little examples like I've done to rediscover the properties whenever you need them.

Instead of trying to memorize the various properties, if you take the time to make sure you can really follow these examples—and even explain them to someone else—then you should have no problems with exponents.

—Dr. Math, The Math Forum

Using Scientific Notation

Dear Dr. Math,

How would you solve a problem using scientific notation? I know it's used to multiply large numbers to get a correct answer, but I don't understand how to do it. Here's what I've tried:

    4,567,839 · 5,493,711 =
    4.567839 · 10 to the sixth power ·
    5.493711 · 10 to the sixth power = ?

I don't know where to go from here. Do I multiply the number by the decimal by the exponent, or ignore the exponent and simply multiply? Please give me an example.

I also understand that you do different things with the problem depending on the operation being used. Help!

From,

Carissa

Hello, Carissa,

The thing to remember when working with **scientific notation** to do multiplication is the law of exponents that says

$$a^m \cdot a^n = a^{(m+n)}$$

For example,

$$10^6 \cdot 10^6 = 10^{(6+6)}$$
$$= 10^{12}$$

Notice that the base has to be the same in both numbers! You cannot apply this law of exponents to $10^6 \cdot 5^6$ because the bases are different—the first number has a base of 10 and the second has a base of 5.

Getting back to your problem, you already know that $4{,}567{,}839 = 4.567839 \cdot 10^6$ and that $5{,}493{,}711 = 5.493711 \cdot 10^6$. This makes the problem $(4.567839 \cdot 10^6) \cdot (5.493711 \cdot 10^6)$.

Applying the Associative Property of Multiplication, which says

$$a \cdot (b \cdot c) = (a \cdot b) \cdot c$$

and the Commutative Property of Multiplication, which says

$$a \cdot b = b \cdot a$$

we get

$$(4.567839 \cdot 5.493711) \cdot (10^6 \cdot 10^6)$$

By multiplying the two left-most factors and applying the above law of exponents to the right-most factors we get

$$25.094387 \cdot 10^{12}$$

So, in a nutshell, we converted our large numbers to scientific notation, added together the exponents where we had common bases, then multiplied the numbers without exponents. You could leave the answer in exponential form or you could expand it to

25,094,387,000,000

The format you choose depends on what you have been asked to use for your answer.

— Dr. Math, The Math Forum

4 Large and Small Numbers

When we deal with numbers in real life, we often talk about things that can only be described using very large numbers or very small numbers. For example, if we are talking about space travel and the distances between the earth and other planets, the numbers we use are quite large. Small numbers are commonly used when we talk about microscopic things. Very large or very small numbers are often most easily expressed using exponents, usually with a technique called scientific notation.

When do you use scientific notation and when do you use exponents? The general answer is that when you have a big number in base 10, you can use scientific notation to make it easier to deal with. But sometimes you don't get your big number in base 10! You get it indirectly from a simulation of some kind of process, like the accumulation of interest, or successive divisions of bacteria, which naturally leads to the use of exponents. For example, if you multiply 186,000 miles per second by 86,400 seconds per day, then multiply by 365.25 days per year, you get 5,869,713,600,000, or $5.87 \cdot 10^{12}$, which is the number of miles in a light-year. But if you let 1,000 dollars accumulate for 120 years at 5 percent interest, you end up with $1{,}000 \cdot 1.05^{120}$ dollars.

Power of 120

Dear Dr. Math,

Is there a quick way to work out 1.05 to the power of 120? The only way I can think of is to take 1.05 and keep multiplying it by itself 120 times.

From,

Clive

Hi, Clive,

I would use a scientific calculator, which has an exponent key. (My computer has one in the Accessories menu.)

If I had to do it by hand, I'd try to make it simpler. Let's look at an example: Say we want to raise 6 to the tenth power. We could multiply 6 by itself 10 times; or we could square it, then multiply that result by itself 5 times; or multiply 6 by itself 5 times and square that result. Here are the possibilities written out:

Notice something cool about the exponents in this example? The prime factors of 10 are 5 and 2.

$$6 \cdot 6 \cdot 6 \cdot 6 \cdot 6 \cdot 6 \cdot 6 \cdot 6 \cdot 6 \cdot 6 = 6^{10}$$
$$(6 \cdot 6) \cdot (6 \cdot 6) \cdot (6 \cdot 6) \cdot (6 \cdot 6) \cdot (6 \cdot 6) = (6^2)^5$$
$$(6 \cdot 6 \cdot 6 \cdot 6 \cdot 6) \cdot (6 \cdot 6 \cdot 6 \cdot 6 \cdot 6) = 6^5 \cdot 6^5 = (6^5)^2$$

So, for your example, I'd try something like

$$1.05 \cdot 1.05 = 1.05^2$$
$$1.05^2 \cdot 1.05^2 = 1.05^4$$
$$1.05^4 \cdot 1.05^4 = 1.05^8$$
$$1.05^8 \cdot 1.05^8 \cdot 1.05^8 = 1.05^{24}$$

Why did I cube 1.05^8 instead of squaring it? Because if I'd squared it, I would have gotten 1.05^{16}, and 16 doesn't divide evenly into 120. But

24 does. If I divide 24 into 120, that tells me how many times I should multiply 1.05^{24} by itself, right? This method takes only 9 multiplications. That seems pretty easy compared to 119!

—Dr. Math, The Math Forum

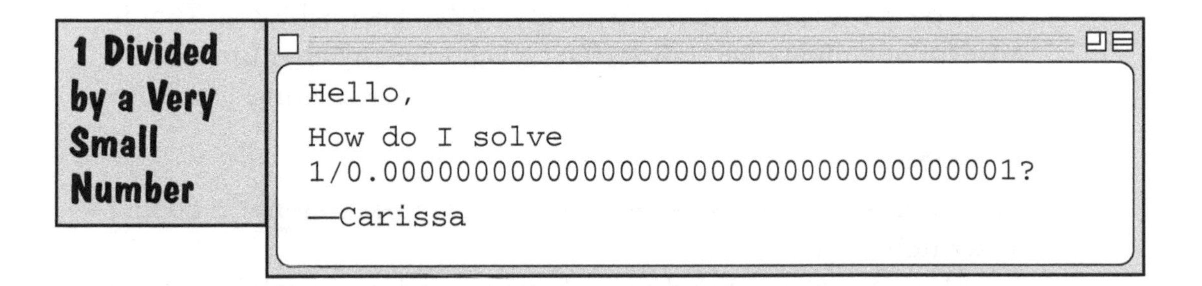
1 Divided by a Very Small Number

Hello,

How do I solve 1/0.00000000000000000000000000000000001?

—Carissa

Dear Carissa,

All you have to do is first count the decimal places. Look at this pattern: $0.1 = 1/10$, $0.01 = 1/10^2$, $0.001 = 1/10^3$, and so on. The exponent that goes with the 10 is equal to the number of decimal places that you counted. The denominator of the above number, 0.00000000000000000000000000000000001, is equal to

$$\frac{1}{10^{35}}$$

which means that your original expression could be written as $\frac{1}{\left(\frac{1}{10^{35}}\right)} = 10^{35}$. In case you are wondering how you get 10^{35}, remember that to divide by a fraction, you can invert and multiply:

$$\begin{aligned}\frac{1}{\left(\frac{1}{10^{35}}\right)} &= 1 \cdot \frac{10^{35}}{1} \\ &= 10^{35}\end{aligned}$$

—Dr. Math, The Math Forum

LARGE NUMBERS AND INFINITY

What is the largest number?

There is no largest number! Why? Well, 1,000,000,000 (1 billion) can't be the largest number because 1 billion + 1 is bigger. That is true for any number you pick. You can choose any big number and someone always can make a bigger one just by adding 1 to it.

What is a googol?

A googol is a 1 with 100 zeros behind it. Using exponents, a googol is written as 10^{100}.

The biggest number that we have named is googolplex—10 to the googol power, or $10^{(10^{100})}$. That's written as 1 followed by a googol zeros!

How do we name large numbers?

There's some disagreement in the English language about how to name large numbers. There are two systems: the American and the British.

American	*Number*	*Scientific Notation*	*British*
Thousand	1,000	10^3	Thousand
Million	1,000,000	10^6	Million
Billion	1,000,000,000	10^9	Thousand million
Trillion	1,000,000,000,000	10^{12}	Billion
Quadrillion	1,000,000,000,000,000	10^{15}	Thousand billion
Quintillion	1,000,000,000,000,000,000	10^{18}	Trillion

There are names for powers of 10 from 1,000 up to as high as you like because some mathematicians have invented systems that you can use to figure out the names for numbers of any size.

It's just that you end up with names like ten duotrigintillion or ten thousand sexdecillion (that's a googol, in American or British, respectively).

But the problem with using names for big numbers is that as soon as you get much above a quintillion, no one is likely to know what you're talking about. So, while it can be fun to try to figure out what the names are, if you're actually trying to communicate with other people, it's better just to use scientific notation.

What is infinity?

Infinity is not a number; it is the name for a concept. Most people have sort of an intuitive idea of what infinity is—it's a quantity that's bigger than any number. This is basically correct, but it depends on the way in which you're using the concept of infinity—some infinities can be larger than others!

It's true that there are no numbers bigger than infinity, but that does not mean that infinity is the biggest number, because it's not a number at all. For the same reason, infinity is neither even nor odd.

The symbol for infinity looks like a number 8 lying on its side: ∞. As soon as they learn about this symbol, some people start sticking it into equations as if it represented a number, which leads them to ask questions like "Which is bigger: $\frac{1}{\infty}$ or $\frac{2}{\infty}$?" or "Why isn't it true that $\frac{1}{0} = \infty$?"

But this just leads to confusion—sort of like if you tried to insert musical notes into a paragraph where words ought to go—so you should resist this temptation.

Now for the fun part! Even though infinity is not a number, it is possible for one infinite set to contain more things than another infinite set. Mathematicians divide infinite sets into two categories: countable and uncountable. In a countably infinite set you can "number" the things you are counting. You can think of the set of natural numbers (numbers like 1, 2, 3, 4, 5, . . .) as being countably infinite. The other type of infinity is uncountable, which means there are too many things to "number." The real numbers are uncountably infinite (numbers like 2.34 . . . and the square root of 2, as well as all integers and rational numbers). In fact, there are more real numbers between 0 and 1 than there are natural numbers (1, 2, 3, 4, . . .) in the whole number line!

Realize that the concept of infinity varies depending on what mathematical subject area you're talking about. If you're simply counting things, then the picture looks something like this:

1, 2, 3, 4, 5, 6, 7, . . .   Countable Infinity   Uncountable Infinity

If you're talking about the number line, then the picture looks something like this:

5 Order of Operations

Operations are methods for combining numbers, like addition, subtraction, multiplication, and division. It's important to know how to carry out these operations. But it's just as important to understand what order to do them in.

In English, the sentence "The boy bit the dog" means something very different from "The dog bit the boy." You need to understand how word order determines meaning in order to be able to communicate with other people.

When mathematicians write $2 + 3 \cdot 4$, they mean that you should multiply 3 by 4 to get 12, then add 2 to get 14. If students read the equation and thought it meant add 2 and 3 to get 5, then multiply by 4, they would get 20, which not the answer the mathematicians intended.

Long ago, mathematicians agreed on a particular order for arithmetic operations, and we've been using it ever since. Some people remember it by PEMDAS, which stands for *p*arentheses, *e*xponents, *m*ultiplication and *d*ivision (left to right), *a*ddition and *s*ubtraction (left to right).

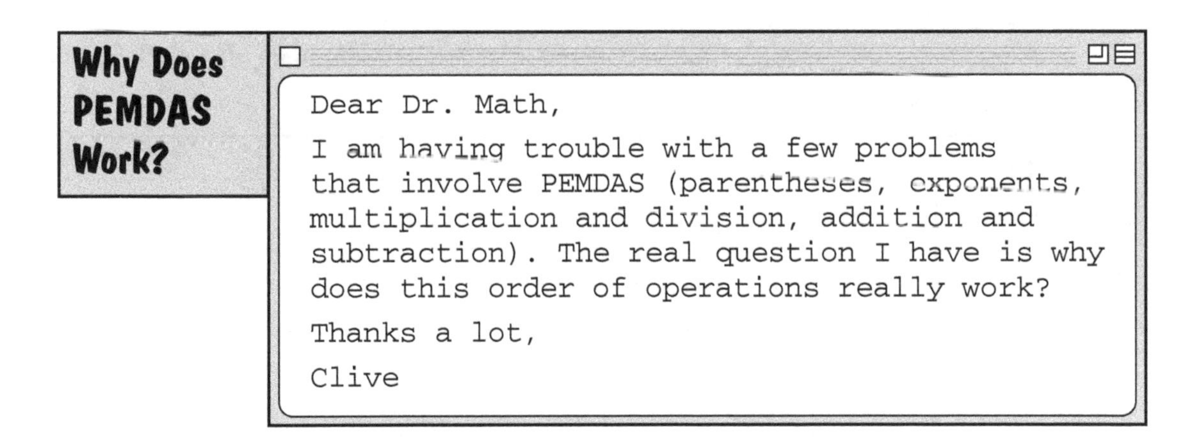

Why Does PEMDAS Work?

Dear Dr. Math,

I am having trouble with a few problems that involve PEMDAS (parentheses, exponents, multiplication and division, addition and subtraction). The real question I have is why does this order of operations really work?

Thanks a lot,

Clive

Hi, Clive,

"Why does this order of operations really work?" does not have a scientific explanation. The order of operations for interpreting a mathematical expression such as $2 + 3 \cdot 5$ is called a convention. A long time ago, people just decided that this was the order in which operations should be performed. It has nothing to do with magic or logic.

It's like asking, "Why is it that when I use the word cow, people tend to think of a beast in a field that moos and gives milk?" A long time ago, someone decided to call that particular type of beast a cow, and now everyone agrees on the use of this word. How efficient! Now people don't have to give long descriptions of grass-chewing, milk-giving, mooing beasts. Instead they can just use the word cow. Like the order of operations, it just has to do with what people have agreed on to make communication with one another a little easier.

So, the answer to your question is that the order of operations works because you are interpreting math formulas the way people have decided to interpret them.

If you're having trouble learning the order of operations, then you have to remember that you're not having a problem with mathematics (unless you're getting wrong answers because you're adding incorrectly). You're having trouble learning a convention, like learning the grammar of a foreign language. Because it's a convention, learning it is just a matter of practice, practice, practice. Eventually you'll have it memorized.

—Dr. Math, The Math Forum

Order of Operations in Equations

Dear Dr. Math,

What is the reason for the order of operations in a math equation? Is it just convention to do multiplication and division before addition and subtraction or is there a deeper reason for this? If there is a deeper reason—and I suspect there is—please include parentheses and exponents in your answer.

Thanks,

Carissa

Hello, Carissa!

The concept of order of operations is really not inherent to the structure of mathematics but rather to mathematical notation—that is, order of operations refers to which operations should be performed in what order, but it doesn't actually dictate anything about (nor is it dictated by) the operations themselves. So, really it's just convention.

If that's true, we should be able to use different orders of operations and come up with a perfectly consistent mathematical system. And, in fact, we can. Here's an example of the same expression being expressed in three different notation systems, resulting in three different orders of operation that all give the same result:

$$5 + 7 \cdot (3 - 2)$$

$$3 - 2 \cdot 7 + 5$$

$$5732 - \cdot +$$

The first expression is the standard one that most people use when writing things down. You do the $3 - 2$ first (because it's in parentheses), then multiply by 7 (because multiplication comes before addition), then add 5.

The second expression is what you would key into a normal calculator (the kind you find at a corner drugstore, not some fancy scientific one with all the extra buttons). You key in the 3, the minus sign, and the 2 first, to subtract. When you hit the times key, it takes the answer to the subtraction and uses that as the first input to the next operation, which is the multiplication by 7. Then when you hit the addition key, it takes the result of all that previous stuff and uses it as the first input to the last operation, which is the addition of 5. (Sounds a lot more complicated than it is, huh?)

The third expression is written in the order in which you'd key it into a fancy scientific calculator like mine. First you give it all the numbers, then you tell it what to do with them. So, the operations say: "subtract the last number in the list from the one before it, then take the result of that and multiply it by the next number back in the list, then take that result and add it to the next number back in the list."

You ask about parentheses and exponents. Parentheses are a way of getting around the order of operations. If I simply wrote $5 + 7 \cdot 3 - 2$, the order of operations says I have to interpret that as 24, because the multiplication comes first. But what if I mean for the subtraction to come first? Then I have to use parentheses: $5 + 7 \cdot (3 - 2)$, which gives me 12 instead of 24.

Exponents are "enhanced multiplication" if you will—a way of multiplying many times quickly—just as multiplication is a way of adding many times quickly. You can think of it like this: the same way multiplication takes precedence over addition in the order of operations, exponents come before multiplication. Or you might think of it this way: the more "powerful" an operation is, the sooner we want to get it out of the way.

We could invent our own order of operations. Let's make one in which adding takes a higher precedence than multiplying. That means subtraction, too, since that's the same as adding a negative. For notation, it means that we don't have to group additions and subtractions in parentheses just so that they get done before multiplications and divisions. Let's take something in our current order of operations, like $(3 + 4) \cdot (7 - 9) \cdot 3 + 6$, and convert it to the new order.

In our current order, this expression tells us to add the 3 to the 4 and subtract the 9 from the 7 first. In our new order, we get that anyway, because addition and subtraction come before multiplication and division. So, we don't need the parentheses that tell us to do these two operations first. Let's take them out:

$$3 + 4 \cdot 7 - 9 \cdot 3 + 6$$

Are we done? Does this match the first expression? Nope, because the next thing we'd do in the new order is add the 3 and 6, and the next thing we should do according to the old order is multiply the whole first part of the expression by 3. How can we get that in our new order? How about using parentheses again?

$$(3 + 4 \cdot 7 - 9 \cdot 3) + 6$$

Now are we done? Let's see: In the new order, we add or subtract first, which gives us

$$(3 + 4 \cdot 7 - 9 \cdot 3) + 6$$
$$(7 \cdot -2 \cdot 3) + 6$$

Then to do our next addition, we have to figure out the quantity in parentheses, so we multiply. I come up with –36 as my final answer. Is that what we would get with the original expression in our original order of operations?

So, the notation tells you which operations to do first, not the underlying mathematics. Of course, there are some pretty good reasons for doing it the normal way: it's much less awkward than the other methods. For instance, how would exponentiation fit into the scheme of the new notation? We switched the order of addition and multiplication, so doesn't it follow that exponents would come last? What if you're adding 3 and 4^2—you can't change the base of the exponent, so how can addition come first?

—Dr. Math, The Math Forum

ASK DR. MATH FAQ

ORDER OF OPERATIONS

To remember the conventional order of operations, you can think of the acronym **PEMDAS** (which you might remember as "Please excuse my dear Aunt Sally"):

1. Parentheses
2. Exponents
3. Multiplication and Division (left to right)
4. Addition and Subtraction (left to right)

This means that you should do what is possible within parentheses first, then exponents, then multiplication and division (from left to right), then addition and subtraction (from left to right).

Instead of PEMDAS, some people are taught to remember **BEDMAS**, which stands for

1. Brackets
2. Exponents
3. Division and Multiplication (left to right)
4. Addition and Subtraction (left to right)

Here are two examples:

$$3 + 5 \cdot 7 = ?$$
$$3 + 5 \cdot 7 = 3 + 35 = 38$$

$$(1 + 3) \cdot (8 - 4) = ?$$
$$(1 + 3) \cdot (8 - 4) = 4 \cdot 4 = 16$$

There are several things that we want to get from a notational system; one of them is consistency: if any two people evaluate the same expression, they should come up with the same answer!

The best reason for using conventional order of operations is the flexibility it gives you in writing down mathematical expressions. Remember that the operations of addition and multiplication are commutative—that is, they give you the same result no matter what order you write them in: $2 \cdot 3 = 3 \cdot 2$. In conventional notation, that property is clearly reflected, and you get lots of options for how you want to write your expressions: $2 \cdot 3 + 5 = 3 \cdot 2 + 5 = 5 + 2 \cdot 3 = 5 + 3 \cdot 2 = 11$. In calculator order, however, you only get to use the first two expressions, $2 \cdot 3 + 5$ and $3 \cdot 2 + 5$, since if you punch the third expression into a calculator, you get 21; and the fourth, you get 16.

6 Distributive Property and Other Properties

If a band of trick-or-treaters comes to your door at Halloween, what do you do with your big bowl of candy? You hand it out, or *distribute* it, evenly to each trick-or-treater. The **distributive property** involves the *distribution* or handing out of a number involved in one operation among two or more numbers involved in another operation. It lets us say

$$3(2 + 5) = 3(2) + 3(5)$$

The 3 was *distributed* to both the 2 and the 5. We call this the *distributive property of multiplication over addition,* and we say that multiplication distributes over addition. There is also the distributive property of multiplication over subtraction, and of exponentiation over multiplication.

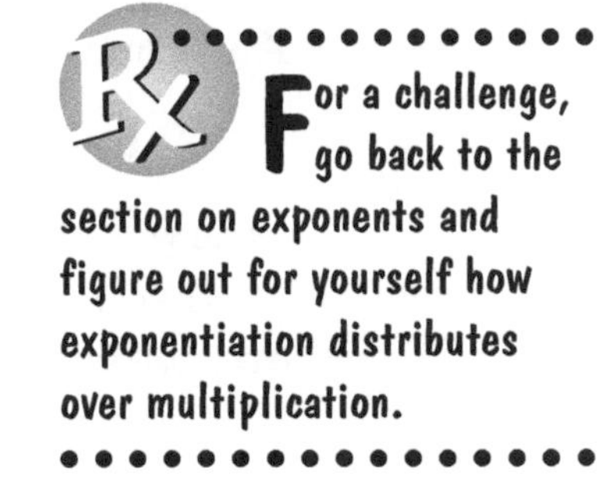

What Is the Distributive Property?

```
Dear Dr. Math,
What is the distributive property?
Clive
```

Hello, Clive,

The distributive property is a pretty cool thing. When we talk about the distributive property, we always talk about the distributive property of one thing over another, like of multiplication over addition or of multiplication over subtraction. If you know what exponents are, you can also talk about the distributive property of exponentiation over multiplication.

Basically, what it says is this: if we have a number times the sum of two numbers, we can *distribute* the first number across the sum and multiply it by each of the second numbers, and then do the adding. For instance, if you have $6 \cdot (4 + 3)$, you can *distribute* the 6 over to the 4 and the 3 to get $6 \cdot 4 + 6 \cdot 3$, and you'll get the same answer you would have gotten if you had done the problem the original way. To check this case, compute the first answer:

$$6 \cdot (4 + 3) = 6 \cdot 7 = 42$$

And compute the second answer:

$$6 \cdot 4 + 6 \cdot 3 = 24 + 18 = 42$$

Whew, it worked!

If you've studied some algebra, here's how the distributive property works out in algebraic symbols: $x(y + z) = xy + xz$. It's something that will come up a whole lot in the future, so you'll want to make sure you're pretty solid on it.

—***Dr. Math, The Math Forum***

Explaining Algebra Concepts

```
My friend Clive needs help with his algebra
and I have tried and tried to help, but he
just doesn't get it. For example, he just
doesn't understand

    4y(5y - 3) + 3y(y + 4)

Thank you,
Carissa
```

Hi, Carissa,

I can appreciate your feeling of frustration. You and I probably see some of these things with no trouble, but other people just need a lot of patient explanation and different ways to look at an idea until some way comes along that clicks. It can be very hard to keep calm and helpful, and not make your friend feel stupid. I'll give you a few ideas that may help prevent frustration.

One thing I like to do is to not only start simple but keep it simple. That means avoiding too many new ideas, and explaining everything in terms of basics. For instance, I don't emphasize memorizing FOIL, because it just looks like one more recipe to memorize, when it's really just a way to keep track of what you multiply when you use the distributive property. FOIL stands for First, Outer, Inner, Last:

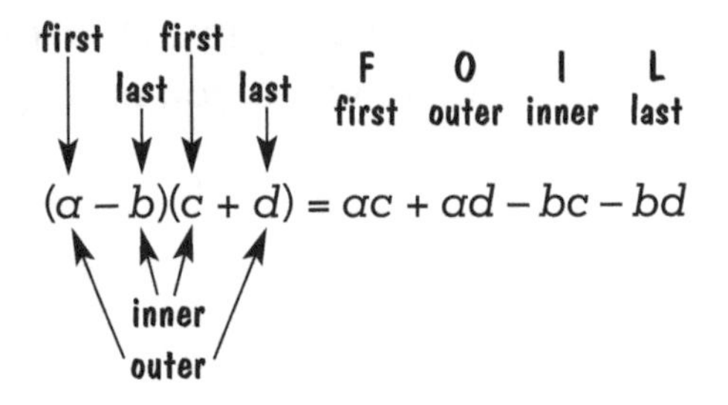

Also, it doesn't apply to multiplying trinomials —for example, $(a - b + c)(d + e + f)$ can't be worked using FOIL, since there's also a "middle" and the pattern breaks down—and can get in the way of a proper understanding.

What I like to do instead is model multiplication of polynomials after multiplication of numbers so that it looks familiar. For example, for $4y(5y - 3)$ I would write

$$\begin{array}{r} 5y - 3 \\ 4y \\ \hline 20y^2 - 12y \end{array} \quad \text{Think of it like regular multiplication:} \quad \begin{array}{r} 73 \\ \times 3 \\ \hline 219 \end{array}$$

After all, what the regular multiplication is saying is

$$\begin{array}{r} 70 + 3 \\ \times 3 \\ \hline 210 + 9 \end{array}$$

This method keeps track of all the "pieces" of each number. It works for larger polynomials, too: for $(a - b)(c + d)$ I would write

$$\begin{array}{r} c + d \\ a - b \\ \hline -bc - bd \\ ac + ad \\ \hline ac + ad - bc - bd \end{array} \quad \text{Compare:} \quad \begin{array}{r} 30 + 7 \\ \times 20 + 2 \\ \hline 60 + 14 \\ 600 + 140 \\ \hline 814 \end{array} \quad \begin{array}{r} 37 \\ \times 22 \\ \hline 74 \\ 74 \\ \hline 814 \end{array}$$

(F O I L)

first, outer, inner, last

This method is most useful when you can group like terms as you go (which is really exactly what you do when you multiply numbers):

$$\begin{array}{r} 3x + 2 \\ 2x - 3 \\ \hline -9x - 6 \\ 6x^2 + 4x \quad\quad \\ \hline 6x^2 - 5x - 6 \\ (F \quad O + I \quad L) \end{array}$$

Do you see how this helps us organize complex multiplications without adding new ideas to worry about? It lets us concentrate on the important idea, which is that multiplication distributes over addition, meaning that each term inside a set of parentheses has to be multiplied by what's outside the parentheses. (For two sets of parentheses, each term in one must be multiplied by each term in the other.) Once you understand distribution thoroughly, it should all fall into place, especially when you have a technique to keep the work organized.

A similar thing happens in solving equations. With simple equations, it can be easy to see the solution, but when there are a lot of things happening in an equation, it can be hard to solve. Your friend may need to learn how to focus on one part of an equation at a time and understand each thing that has to be done. I like to think of this as peeling an onion: you have to do it one layer at a time. When you solve a complicated equation, you have to look for the "outermost" part of the expression and pull that off, ignoring the rest. For example, say you have to solve

$$6(3(4x - 2) - 4) + 3 = 0$$

The parentheses protect the inner part of the equation, so before we get to that part, we have to remove the 3 (by adding –3 to both sides), then the 6 (by dividing both sides by 3):

$$\begin{aligned} 6(3(4x - 2) - 4) &= -3 \\ 3(4x - 2) - 4 &= -3/6 \end{aligned}$$

Now we can work on the next layer:

$3(4x - 2) - 4 = -1/2$	Simplify the right side.
$3(4x - 2) = 7/2$	Add 4 to each side.
$4x - 2 = 7/6$	Divide by 3.
$4x = 19/6$	Add 2 to each side.
$x = 19/24$	Divide by 4.

Of course, you can also work this sort of problem from the inside out, simplifying the expression so that there aren't so many layers, before you start peeling off what's left. If we did that, our example would look like this:

$6(3(4x - 2) - 4) + 3 = 0$	Multiply the inner parentheses by 3.
$6((12x - 6) - 4) + 3 = 0$	Subtract 4.
$6(12x - 10) + 3 = 0$	Multiply the parentheses by 6.
$(72x - 60) + 3 = 0$	Add 3.
$72x - 57 = 0$	Add 57 to both sides.
$72x = 57$	Divide by 72.
$x = 57/72$	Simplify (3 goes into both parts of the fraction).
$x = 19/24$	

—Dr. Math, The Math Forum

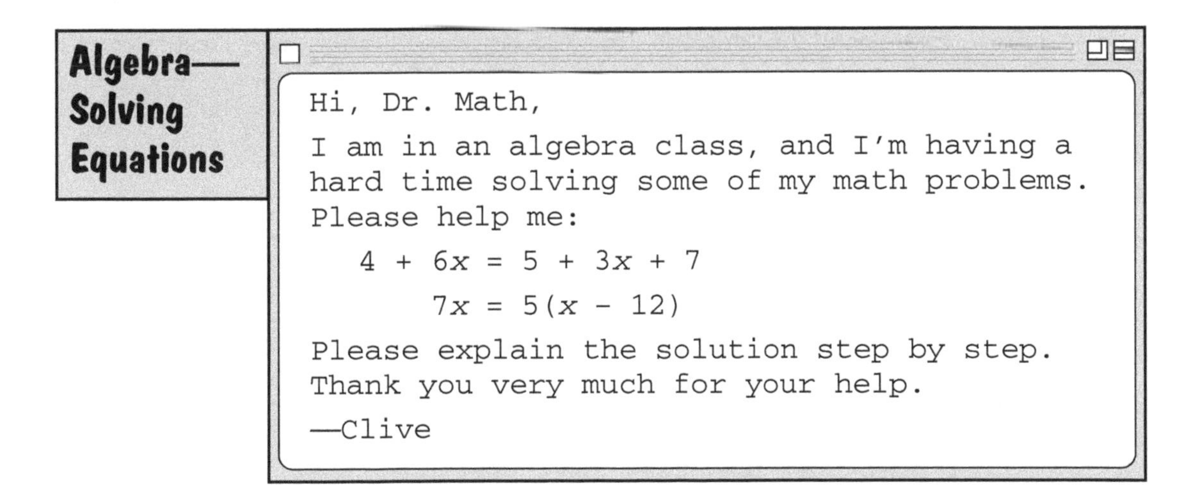

Algebra—Solving Equations

Hi, Dr. Math,

I am in an algebra class, and I'm having a hard time solving some of my math problems. Please help me:

$$4 + 6x = 5 + 3x + 7$$

$$7x = 5(x - 12)$$

Please explain the solution step by step. Thank you very much for your help.

—Clive

Hello, Clive,

I will give you techniques to use when solving these equations, then help you with the first of your problems. You can check out whether that helps you to solve the second problem and get the general picture.

When you solve an equation, you want to find values such that if you substitute these values for x in your equation, you will get something that is true. Let me give you an example:

$$x + 5 = 8$$

You need to find a value to substitute for x such that you get 8. If you have 5 apples and you want to have 8 apples, how many more apples should you buy?

That's basically what an equation is all about. So, the smart thing to do is to isolate x on one side of the equation, and that will tell you the exact value to be substituted. In order to do this, you have to remember some rules. The most important one is that whatever you do to one side of the equation, you should do to the other side. For instance, if you add 3x to one side, you should add 3x to the other side. You should also remember the distributive property:

$$x(a + b) = (a + b)x = ax + bx$$

When you multiply terms within parentheses by x, you multiply x by each term in the parentheses.

That should cover the basics. Now let's go to the first of your problems:

$$4 + 6x = 5 + 3x + 7$$

First notice that $5 + 7 = 12$, so you get

$$4 + 6x = 12 + 3x$$

Now you want to isolate x on one side of the equation, so let's get rid of the 4 on the left side of the equation by subtracting 4 from each side of the equation. That gives you

$$4 + 6x - 4 = 12 + 3x - 4$$

or, simplifying,

$$6x = 8 + 3x$$

Since you want the x to be only on one side, get rid of the 3x on the right side of the equation by subtracting 3x from both sides:

$$6x - 3x = 8 + 3x - 3x$$

which is

$$3x = 8$$

All you need to do now to isolate x is divide each side by 3:

$$\frac{3x}{3} = \frac{8}{3}$$

So, you get

$$x = \frac{8}{3}$$

—***Dr. Math, The Math Forum***

ASK DR. MATH FAQ

GLOSSARY OF PROPERTIES

Operation

An *operation* works to change numbers. (The word *operate* comes from the Latin *operari,* "to work.") There are five operations in arithmetic that work on numbers: addition, subtraction, multiplication, division, and raising to powers. (Remember that raising a number to a fractional power is the same as taking a root of that number; e.g., $x^{\frac{1}{2}}$ is the same as $\sqrt{x}$.)

Identity

An *identity* is a special kind of number. When you use an operation to combine an identity with another number, that number stays the same. Zero is called the *additive identity,* because adding zero to a number will not change it: the number stays the same.

$$0 + a = a = a + 0$$

Since any number multiplied by 1 remains constant, the *multiplicative identity* is 1.

$$1 \cdot a = a = a \cdot 1$$

Inverse

The inverse of something is that thing turned inside out or upside down. The inverse of an operation undoes the operation: division undoes multiplication.

A number's *additive inverse* is another number that you can add to the original number to get the additive identity. For example, the additive inverse of 67 is –67, because $67 + -67 = 0$, the additive identity. Zero's additive inverse is itself.

Similarly, if the product of two numbers is the *multiplicative identity,* the numbers are *multiplicative inverses.* Since $6 \cdot \frac{1}{6} = 1$ (the multiplicative identity), the multiplicative inverse of 6 is $\frac{1}{6}$.

Zero does not have a multiplicative inverse, since no matter what you multiply it by, the answer is always 0, not 1.

Associative property

An operation is associative if you can *group* numbers in any way without changing the answer. It doesn't matter how you combine them, the answer will always be the same. Addition and multiplication are both associative. Here are some addition examples:

$$1 + (2 + 3) = (1 + 2) + 3$$
$$1 + (5) = (3) + 3$$
$$6 = 6$$

$$(-1 + 66) + 14 = -1 + (66 + 14)$$
$$(65) + 14 = -1 + (80)$$
$$79 = 79$$

In general, we can say that

$$a + (b + c) = (a + b) + c$$

Here is a multiplication example:

$$2 \cdot (4 \cdot 3) = (2 \cdot 4) \cdot 3$$
$$2 \cdot (12) = (8) \cdot 3$$
$$24 = 24$$

In general, we can say that

$$a \cdot (b \cdot c) = (a \cdot b) \cdot c$$

Subtraction isn't associative: $5 - (4 - 3)$ does not equal $(5 - 4) - 3$. And neither is division: $(\frac{3}{4})/5$ does not equal $3/(\frac{4}{5})$.

Commutative property

An operation is commutative if you can *change the order* of the numbers involved without changing the result. Addition and multiplication are both commutative. Subtraction is not commutative: $2 - 1$ is not equal to $1 - 2$. Neither is division: $\frac{2}{3}$ is not equal to $\frac{3}{2}$. Here are some examples of the commutative properties of addition and multiplication:

$$88 + 65 = 65 + 88$$
$$153 = 153$$

$$12 \cdot 13 = 13 \cdot 12$$
$$156 = 156$$

In general, we can say that

$$a + b = b + a$$

and

$$a \cdot b = b \cdot a$$

Distributive property

When you distribute something, you give pieces of it to different people. One example of distributing objects is handing out papers in class. In math, people usually talk about the distributive property of one operation over another.

The most common distributive property is the distribution of multiplication over addition. It says that when a number is multiplied by the sum of two other numbers, the first number can be distributed to and then multiplied by both of those two numbers separately. Here's the distributive property in variables:

$$a \cdot (b + c) = a \cdot b + a \cdot c$$

For example,

$$\begin{aligned} 5 \cdot (2 + 8) &= 5 \cdot 2 + 5 \cdot 8 \\ 5 \cdot (10) &= 10 + 40 \\ 50 &= 50 \end{aligned}$$

Not all operations are distributive. For instance, you cannot distribute exponentiation over addition. Let's try an example:

$$\begin{aligned} &(2 + 3)^4 \\ &= 5^4 \\ &= 625 \end{aligned}$$

but

$$\begin{aligned} &(2 + 3)^4 \\ &= 2^4 + 3^4 \quad \leftarrow \text{Don't try this at home! It's not true!} \\ &= 16 + 81 \\ &= 97 \end{aligned}$$

Clearly, since 97 is not equal to 625, exponentiation cannot be distributed over addition. Remember, though, that it *does* distribute over multiplication:

$$\begin{aligned} (2 \cdot 3)^4 &= 2^4 \cdot 3^4 \\ 6^4 &= 16 \cdot 81 \\ 1296 &= 1296 \end{aligned}$$

Resources on the Web

Learn more about fundamental operations at these Math Forum sites:

Algebraic Problem Solving Using Spreadsheets

mathforum.org/workshops/sum98/participants/sinclair/problem/intro.html

A unit that emphasizes moving from numerical to symbolic representation.

ESCOT Problems: Galactic Exchange

mathforum.org/escotpow/print_puzzler.ehtml?puzzle=31

mathforum.org/escot/galactic.html

Students are asked to discover the exchange rates among different types of alien currency and to use this information to find out the amount of money needed to buy certain types of food.

Leonardo da Vinci

mathforum.org/alejandre/frisbie/math/leonardo.html

Question: Is the ratio of our arm span to our height really equal to 1? If an emphasis is given to $a/h = 1$, this can be an engaging activity using variables.

Locker Problem

mathforum.org/alejandre/frisbie/locker.html

A classroom activity (also called 1000 Lockers) to be explored through the use of manipulatives and a ClarisWorks spreadsheet. Students look for patterns and write the answer algebraically.

Middle School Algebra Links

mathforum.org/alejandre/frisbie/math/algebra.html

A variety of resources on the Web emphasizing algebraic thought while addressing the NCTM Standards for grades 6, 7, and 8.

Integers

As you learn about different sets of numbers, it helps to think about when certain groups are needed. When we are simply counting, we use the **natural numbers,** or **counting numbers,** such as 1, 2, 3, ... When we start combining numbers by adding, the natural numbers are still enough, but we run into a little trouble when we need to subtract. For example, if we subtract 3 – 3, we need to have a zero. When we have a problem like 3 – 9, we need to have negative whole numbers in our collection. **Integers** are the set of numbers that includes positive whole numbers, negative whole numbers, and zero.

Here we're going to expand the set of numbers that we deal with by adding zero and the negative numbers to the counting (or positive) numbers. We'll examine the way that these numbers behave when you use them with the operations you've seen before, like addition and multiplication.

In this part, Dr. Math explains

- Coordinate graphing of integers
- Origin of integers
- Adding and subtracting integers
- Multiplying and dividing integers
- Absolute value

1 Coordinate Graphing of Integers

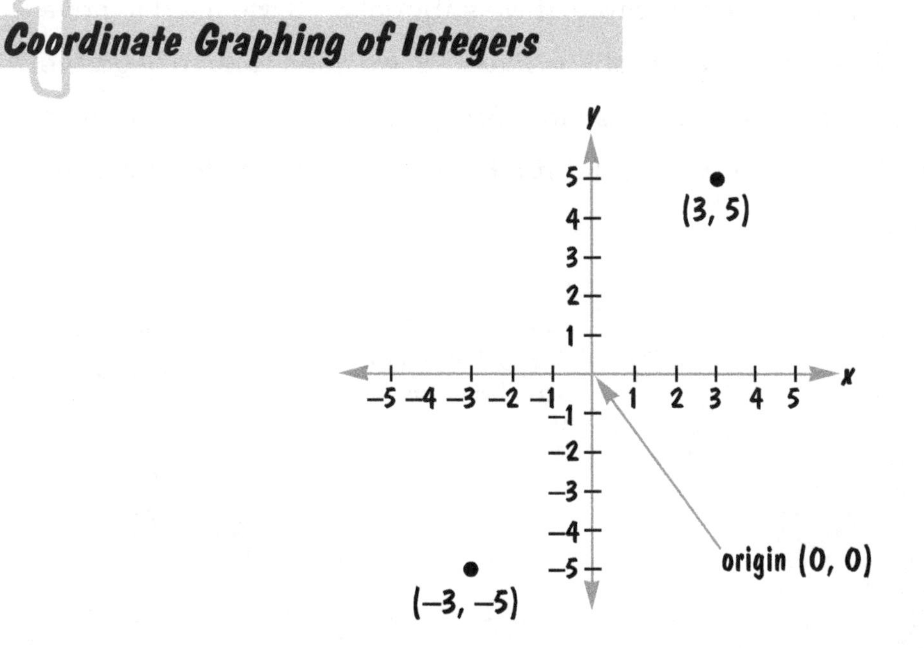

When we **graph** a **coordinate** pair such as (3, 5), we start at the origin (0, 0), go right 3 spaces, go up 5 spaces, and mark the point. When we graph positive numbers, we only have to work with up and right. If we're graphing integers, though, we have to know what to do if one of the coordinates is a negative number. This is where left and down come in.

When we graph a coordinate pair such as (–3, –5), we still start at the origin, but this time we go *left* 3 spaces, *down* 5 spaces, and mark the point. When graphing, we normally define up and right as normal or **positive,** and down and left as **negative.**

Plotting Coordinates on a Graph

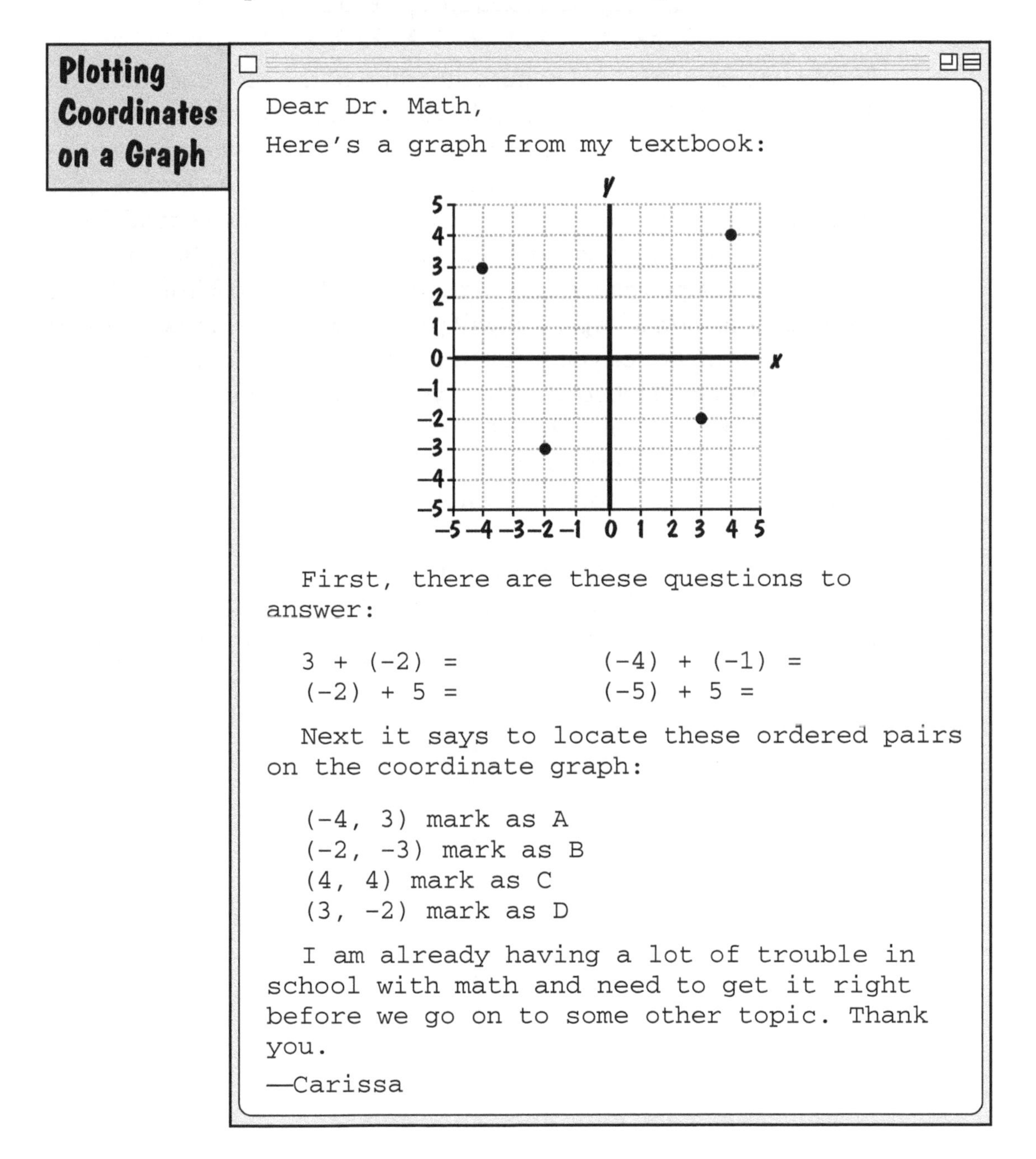

Dear Dr. Math,

Here's a graph from my textbook:

First, there are these questions to answer:

3 + (-2) = (-4) + (-1) =

(-2) + 5 = (-5) + 5 =

Next it says to locate these ordered pairs on the coordinate graph:

(-4, 3) mark as A
(-2, -3) mark as B
(4, 4) mark as C
(3, -2) mark as D

I am already having a lot of trouble in school with math and need to get it right before we go on to some other topic. Thank you.

—Carissa

Hi, Carissa,

The first set of questions is about positive and negative numbers on the number line:

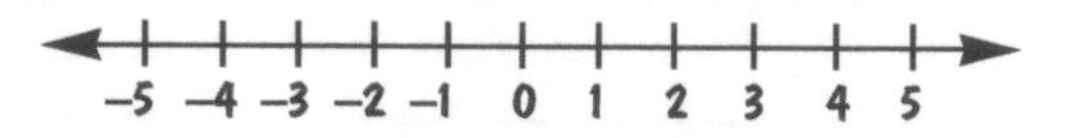

To add two numbers, just picture standing at the zero and facing to the right. A positive number tells you to walk forward that many steps, and a negative number means to walk backward. For example, 3 + –2 means walk forward 3 steps (to the 3), then backward 2 steps. The first step backward takes you to 2, and the second takes you to 1, so that's the answer. What you're really doing is just undoing 2 of the 3 steps you took to the right or subtracting 2 from 3. If you go back more than 3 steps, you will pass zero and be in the negative numbers; the distance you would end up from zero would be the difference between the two distances you walked. For example, 3 + –5 is –2, because the first 3 steps backward take you to zero, and you have 5 – 3 = 2 more steps to take.

See how well you can work out the rest of the problems thinking this way.

The other set of problems relates to this graph, where we use pairs of numbers called **ordered pairs** to identify points on a pair of **axes,** just as we use one number to name points on the number line:

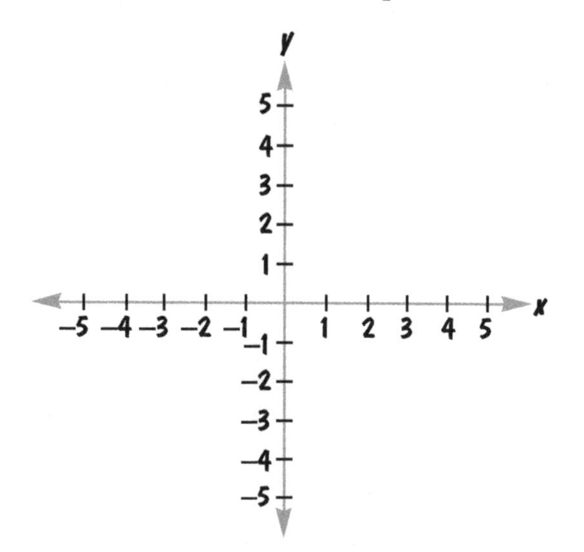

Point A is (–4, 3), which means you want x to be –4 and y to be 3. The distance left or right from the vertical line is x, so we can draw a line vertically through –4 on the x-axis, marking all the points that are 4 units left of the y-axis:

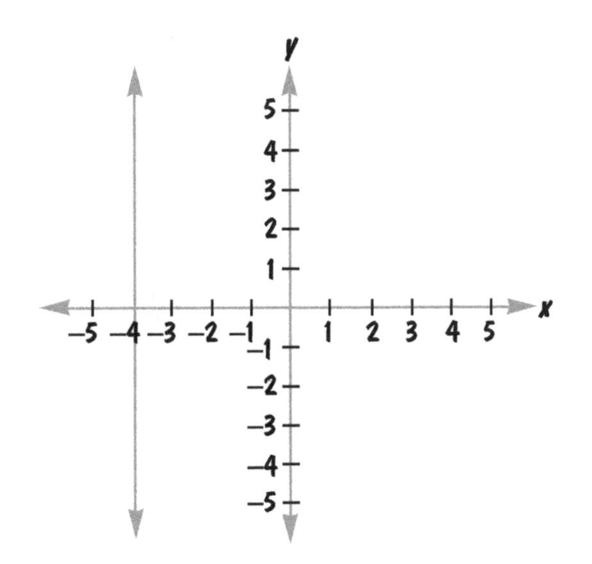

The distance up or down from the horizontal line is y, so we can draw another line through 3 on the y axis:

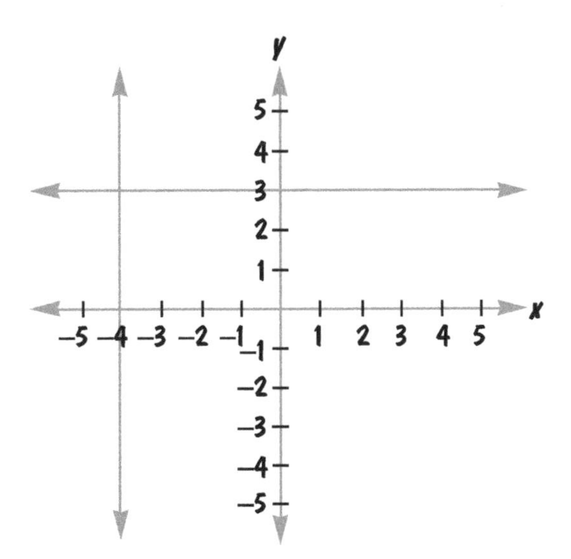

The place where these two lines intersect is A, because it's 4 units to the left and 3 units up. The coordinates (–4, 3) mean "walk 4 steps to the left, then 3 steps up."

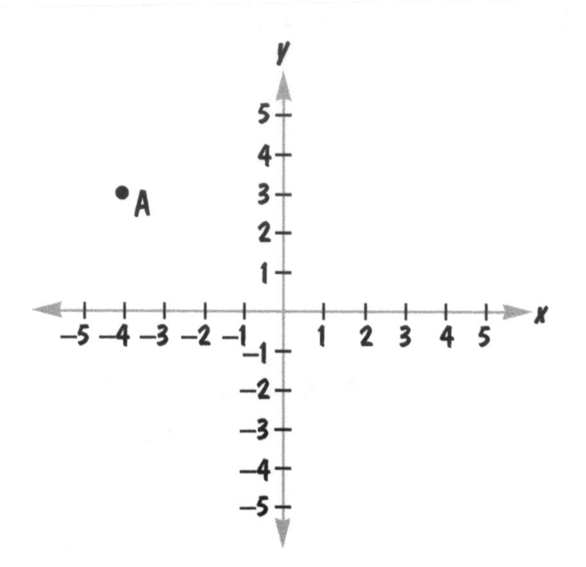

If you follow this method for the other points, your final graph, with all points plotted, will look like this:

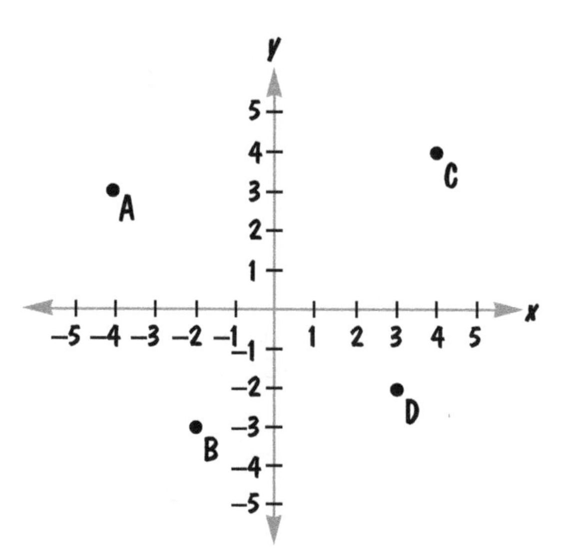

Mathematicians and teachers did not make up coordinates to torture their students. Suppose we're standing near some railroad tracks and I tell you that someone needs help a mile away. It's pretty important to know the exact direction! How might we specify that? I could say, "He's a mile to the left." But if you're on the opposite side of the tracks facing me, do I mean your left or my left? And what if we both step across the tracks? Our "lefts" change. But I can say, "He's a mile to the east," and that description holds no matter which side of the tracks we're on and which way we're facing. That's because we've established a convention that anyone can use to tell which way is east.

In two dimensions, we need two directions. And, in any case, we need a convention to keep people from misunderstanding one another. The convention we've adopted is that we'll use signs instead of writing right, left, up, and down; that we'll specify the horizontal direction before the vertical direction; and that we'll make right and up the default directions. There's nothing special about these choices. The first people to start using these conventions could have chosen different ones. But they didn't, so now we have to learn the same conventions in order to make sense of what they wrote . . . and in order to make sense of one another now. This will be covered in a little more detail in Section 2 of this part.

—Dr. Math, The Math Forum

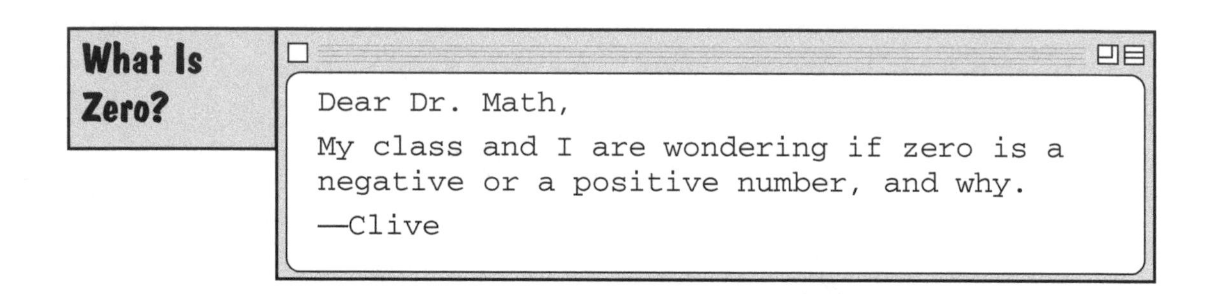

What Is Zero?

Dear Dr. Math,

My class and I are wondering if zero is a negative or a positive number, and why.

—Clive

Dear Clive,

Actually, zero is neither a negative nor a positive number. The whole idea of positive and negative is defined in terms of zero. Negative numbers are numbers that are smaller than zero, and positive numbers are numbers that are bigger than zero. Since zero isn't bigger or smaller than itself (just like you're not older than yourself or taller than yourself), zero is neither positive nor negative.

People sometimes talk about the "nonnegative" numbers, meaning all the numbers that aren't negative—in other words, all the positive numbers and zero. The only difference between the set of positive numbers and the set of nonnegative numbers is that zero isn't in the first set, but it is in the second. Similarly, the "nonpositive" numbers are the negative numbers together with zero.

—Dr. Math, The Math Forum

Origin of Integers

Part of understanding what you can *do* with integers (adding, subtracting, multiplying, dividing) is knowing why we have them in the first place. Historically, number systems have developed because of a need to extend some pattern we've found useful with a smaller or different set of numbers. This section explains why the concept of integers started and why knowing about them is useful.

Why Integers?

```
Dear Dr. Math,
Where did integers come from?
—Clive
```

Hi, Clive,

Part of learning more about math is learning about different kinds of numbers. But why do we need all of them? Where do they come from? What are they good for?

The counting numbers (1, 2, 3, . . .) are useful for counting things like how many sheep are in a flock or how many people are expected for dinner—and you already know how to add and subtract them. If we start out with 10 sheep and wolves eat 2 of them, we have 10 – 2 = 8 sheep. In this context, we can look at an operation like

8 sheep – 10 sheep = ?

and conclude that it simply doesn't make sense. But as we start to deal with things that are more abstract than sheep, it becomes easier to think of contexts in which subtracting a larger number from a smaller one starts to make sense. For example, if a business has $1 million in its bank accounts, it can still spend more than $1 million on equipment or raw materials or advertising—by borrowing money!

A business that starts with $1 million and spends $2 million has gone from having $1 million in credit to having $1 million in debt.

Now our counting numbers are no longer sufficient to tell us what's going on. If we say that Clivissa, Inc., has $1 million, it makes a big difference whether we're talking about credit or debt!

We could just be very careful to always attach a word like *credit* or *debt* to make it clear what we're talking about, which would let us write equations like

$1 million credit – $2 million credit = $1 million debt

But this is a lot of extra writing, so it makes sense to think about ways to make our notation more concise. And we've settled on a notation that is very concise! We attach a minus sign to a number to indicate that we're talking about a quantity that is *opposite* to whatever the normal sense of a number is. We call a number with this minus sign attached to it a *negative* number (although perhaps *opposite* number would be a better name). So, if we think of positive numbers as credit, we can simply write

$1 million – $2 million = –$1 million

which has exactly the same meaning as

$1 million credit – $2 million credit = $1 million debt

The only difference is that we're using a single notation—the minus sign—to stand for all the different kinds of opposites that we'll ever want to talk about using numbers. If "5 miles" means 5 miles east, then "–5 miles" means 5 miles west. In contrast, if "5 miles" means 5 miles north, then "–5 miles" means 5 miles south.

Now we can abbreviate

Drive 5 miles north

Drive 2 miles south

Drive 6 miles north

Drive 14 miles south

Where do you end up?

as

$5 + -2 + 6 + -14 = ?$

On the one hand, this can cause a lot of confusion if you lose track of what you've defined as conventional! On the other hand, it means that we can just develop a single set of rules for dealing with opposites instead of one set of rules for money, one for driving directions, one for airplane instruments, and so on.

This is important: You can save yourself an enormous amount of frustration in learning about negative numbers if you understand that there is nothing that we can do with negative numbers that we can't do without them! The only thing they allow us to do is to stop writing things like *credit* and *west* and *up* all over the place. But, as you get into algebra and start writing lines and lines of equations, you'll see that this one little benefit actually turns out to be very big!

In the end, you do have to learn a few rules that are confusing at first, but they'll seem completely natural after you've used them for a while.

—Dr. Math, The Math Forum

Adding and Subtracting Integers

We already know how to add and subtract positive integers. And we talked briefly in the first section about what happens when we include negative integers in our problems. But all of those minus signs can get confusing. In this section, we will give you some ideas on how to keep everything straight.

Introduction to Negative Numbers

```
Dear Dr. Math,
What does 11 - 12 equal?
—Carissa
```

Hi, Carissa,

That's a good question. It's easy to say what 12 – 11 is, so shouldn't 11 – 12 have an answer, too? Until recently, my own daughter would have written 11 – 12 = C, which stands for "Can't do it."

For some questions, that's the best answer you can give. Suppose you have 11 carrots on your dinner plate, and because I love carrots so much, while your head is turned I try to take 12 carrots from your plate. I can't do it!

But suppose you have $11 in your bank account and you want to buy a game that costs $12. You can't do it, but I might be kind enough to lend you $1. Then you would spend all $11 of your money and owe me $1. You would actually have $1 less than nothing, because as soon as you earned another dollar, it would go to me and you would have nothing!

Several hundred years ago, mathematicians realized that there were a lot of problems they could solve if they had a way to talk about numbers less than zero. They decided to write $0 - 1 = -1$, which is read as "negative one" and means "one less than zero." From what I just said about owing $1, you can see that

$11 - 12 = -1$ (When you spend $12, you owe me $1.)

and

$-1 + 1 = 0$ (When you earn another dollar, you will have nothing.)

There's actually another place where you can see negative numbers very easily. Look at a thermometer and you'll see that the temperature can go down to zero degrees, but then it can keep getting lower. Temperatures below zero are written as negative numbers, like

–10 degrees

which means 10 degrees below zero. If the temperature now is 11 degrees and it gets 12 degrees colder, it will be –1 degree.

—Dr. Math, The Math Forum

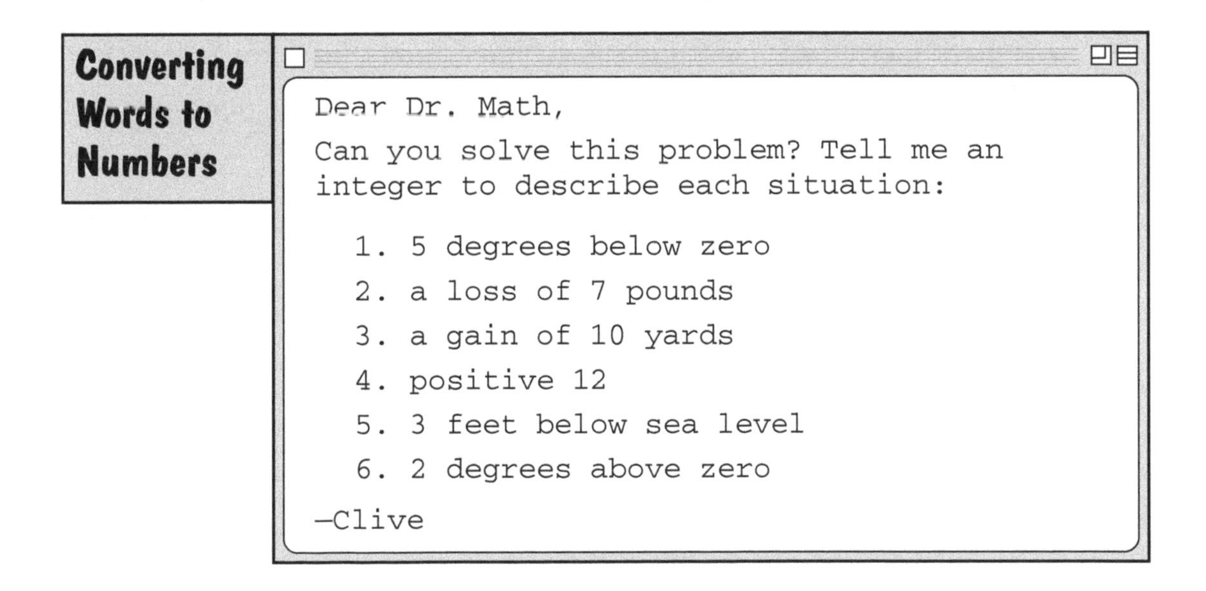

Converting Words to Numbers

Dear Dr. Math,

Can you solve this problem? Tell me an integer to describe each situation:

1. 5 degrees below zero
2. a loss of 7 pounds
3. a gain of 10 yards
4. positive 12
5. 3 feet below sea level
6. 2 degrees above zero

—Clive

Hi, Clive,

Your set of questions deals with conventions—that is, people who use math have generally agreed that certain directions should be conventionally thought of as positive. Anything that is *above* or *up* or *forward* or *increasing* or *more* is positive, and anything that is *below* or *down* or *backward* or *decreasing* or *less* is negative.

For example, if I make a profit of $5 in my business, I would call that +5, and if I lose $5, that would be –5. Why? Because whichever happens, I can add that number to my bank account to find out how much I have now.

Similarly, if a mountain's base is 2 miles below sea level and its peak is 3 miles above sea level, then the altitude of its base is –2 and the altitude of its peak is +3, so the total height is (+3) – (–2) = 5 miles.

As I said, these are just conventions, and they really depend on what you're measuring. If I were in a submarine measuring depth, I would say my depth is +2 miles, because when I think about *depth* I mean something that increases as I go deeper. So a depth of +2 means the same thing as an altitude of –2.

Therefore, the answer to these questions should include some sort of label. For instance, I would say "altitude = –3" for problem 5. You could also say "depth = +3" if you want to confuse your teacher, but then you'd have to bring me in to testify on your behalf, so maybe you'd better stick with –3.

You should be able to do the rest by looking for words like *below* or *loss* to indicate a negative number.

—Dr. Math, The Math Forum

Subtracting Negative Numbers

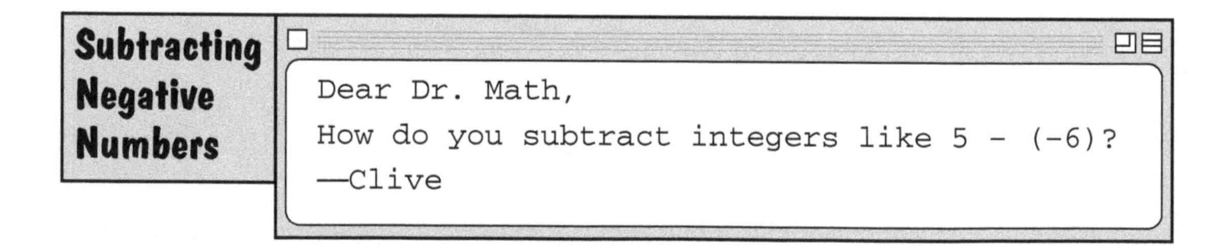

Dear Dr. Math,

How do you subtract integers like 5 - (-6)?

—Clive

Dear Clive,

Let's use the number line to visualize how to find the answer to your problem:

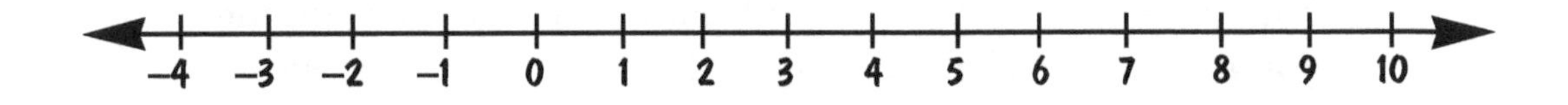

When you add two positive numbers on the number line, you start at one of the numbers and move to the right from that number a distance equal to the other number.

Think about the sum of 5 + 2. From the number 5, we have to hop 2 places to the right. We hop 2 places to the right because we are adding 2 more places.

For the answer to 5 + 2, we will land on the number 7.

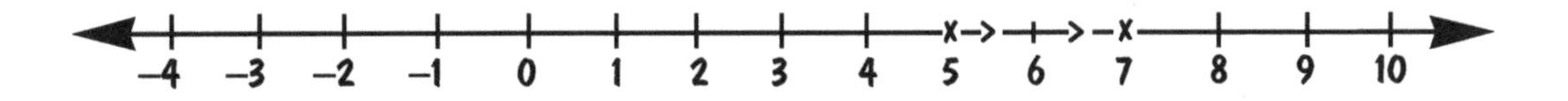

So, 5 + 2 = 7.

Now let's think about subtraction. How would you solve the problem 5 – 2 = ? From the number 5, you would hop 2 places to the left and land on the number 3.

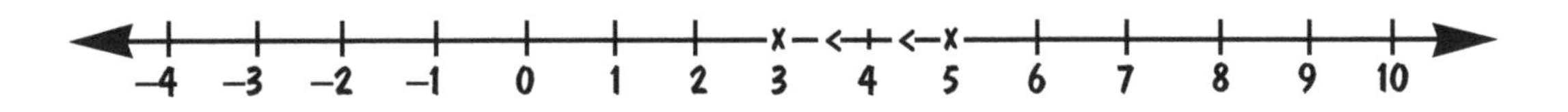

So, 5 – 2 = 3.

To add two numbers, you start at one of the numbers and hop to the right according to the value of the other number. To subtract two numbers, you start at the first number and move to the left according to the value of the second number.

But you need to know what to do to subtract a *negative* number. Once you know which way to go on the number line, it's easier. For instance, consider the following problem:

$$5 - (-2) = ?$$

You know that when you subtract a number from another number, you hop to the left according to the value of the second number. So, to figure out what $5 - (-2)$ is, we would hop to the left -2 hops. But what does that mean?

Since the negative sign reverses everything, moving to the left -2 hops really means that you should move to the right $+2$ hops! Subtracting negative numbers, therefore, is the same as adding the numbers. So, $5 - (-2) = 5 + 2$.

From number 5, when you hop 2 places forward, you will land on the number 7. So, $5 - (-2) = 7$.

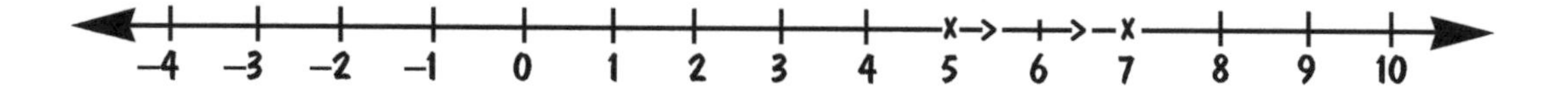

For the sum of $5 - (-6)$, I am sure you have the idea by now.

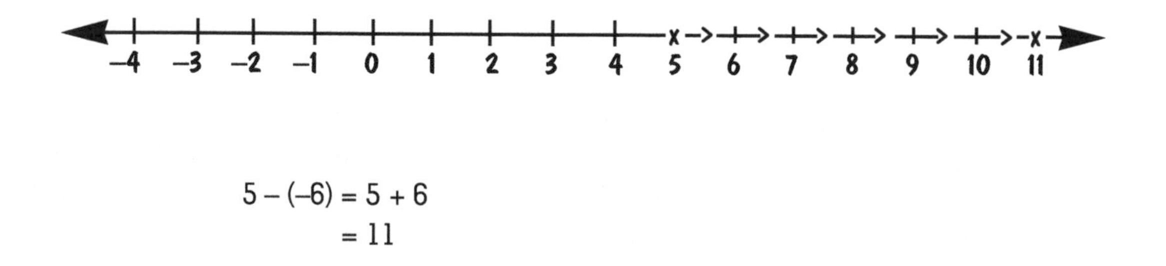

$$5 - (-6) = 5 + 6$$
$$= 11$$

Once you understand the basics and practice them often enough, these types of problems will be easy.

—Dr. Math, The Math Forum

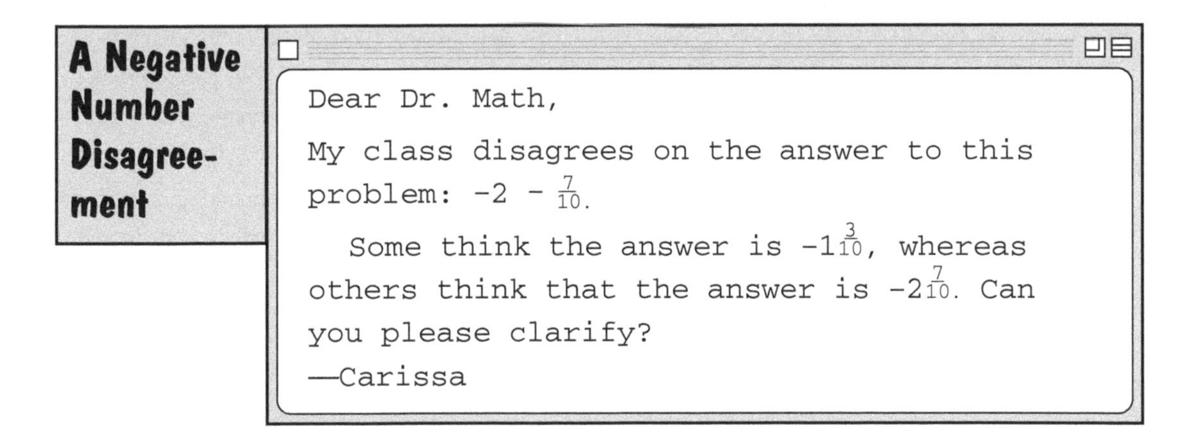

A Negative Number Disagreement

Dear Dr. Math,

My class disagrees on the answer to this problem: $-2 - \frac{7}{10}$.

Some think the answer is $-1\frac{3}{10}$, whereas others think that the answer is $-2\frac{7}{10}$. Can you please clarify?

—Carissa

Hi, Carissa,

Well, the answer is $-2\frac{7}{10}$, but that's more of a convention than anything else. As you know, when you write a mixed number, like $1\frac{2}{3}$, it means to *add* the integer (1) and the fraction ($\frac{2}{3}$), so it is really the same thing as saying $1 + (\frac{2}{3})$.

However, $A\frac{b}{c}$ (i.e., "*A* and *b* cths") does *not* always mean add the integer *(A)* and the fraction ($\frac{b}{c}$), and that's where the confusion lies. It only means add the integer and the fraction *when there is* no *minus sign to the left of the mixed number.*

If you put a negative sign in front of (to the left of) a mixed number, that must mean to take the negative of the entire mixed number. When you write $-2\frac{7}{10}$, you mean the negative of the number $2\frac{7}{10}$—that is, you mean

$$-2\frac{7}{10} = -(2\frac{7}{10}) = -(2 + \frac{7}{10}) = -2 - \frac{7}{10}$$

So, in this case, writing the fraction ($\frac{7}{10}$) next to the number (-2) actually means subtract, not add.

The fractional part of a mixed number works like the decimal part of a number, like -1.25, which can also be written as $-1\frac{1}{4}$. We can't treat them differently in this situation, because they mean the same thing. You could see this in a similar way by changing the mixed number into an improper fraction: $-\frac{5}{4}$. It's pretty clear that we can't use $-\frac{5}{4}$ to mean $-\frac{3}{4}$!

Because of difficulties like this, we have to agree that putting a minus sign in front of a mixed number means take the negative of the *entire* mixed number, *not* add the (positive) fraction part and the negative integer part. However, I would say that the mistake is an extremely subtle one, and I think it would have caught a lot of people by surprise. If you were in my class and you made a mistake like that, I would have told you that this was an extremely smart mistake to make and that you were probably a pretty smart kid to make it!

—Dr. Math, The Math Forum

4 Multiplying and Dividing Integers

About the time that you understand how to add and subtract integers, the subject of multiplying and dividing integers will come up. Reviewing and thoroughly understanding the operations using whole numbers will help you understand how things work with integers. With addition and subtraction, a number line is a good tool. Picturing what is going on when multiplying or dividing integers is not easy. There isn't just one visualizing method that works for everyone, so we have tried to include a variety of ideas to help you understand how to multiply and divide integers.

Working with Integers

Dear Dr. Math,

I'm having trouble figuring out how to multiply and divide integers. Could you explain to me how to do integers? Maybe include some examples? It would really help!

—Carissa

Hi, Carissa,

Integers are simply counting numbers with a sign attached, plus zero to keep the positives and negatives from mixing. The sign tells you which way to go on the number line: plus means to the right; minus means to the left. I'll call the number part the *absolute value* of the integer. For example, the integer –3 consists of the sign "–" and the absolute value 3. If there's no sign, it means "+." We only write "+" when we want to emphasize that the number is a positive integer.

Multiplication and division of integers both follow the same rule: you multiply the absolute values together and choose a sign for the answer based on the combination of signs attached to the absolute values using this table:

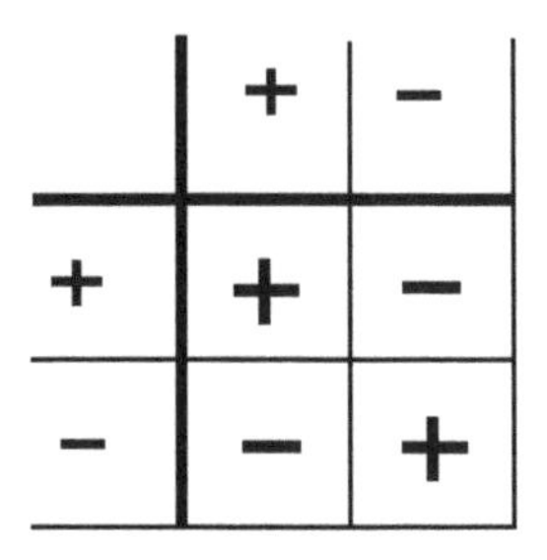

	+	–
+	+	–
–	–	+

When you multiply or divide "+" by "+", or "–" by "–" (the same sign), the answer is "+." If the signs are "+" and "–", or "–" and "+" (different signs), the answer is "–". You can think of "–" as flipping the whole number line over, so if you flip it twice in a row, everything is back where it started: "– –" = "+." When you have more than two integers to multiply or divide, deal with just two at a time to get your answer. This works because $-x = -1x$. So, you could write $-2 \cdot 3$ as $(-1 \cdot 2) \cdot (1 \cdot 3)$, which is $(-1 \cdot 1) \cdot (2 \cdot 3)$, which is $-1 \cdot 6$, which is -6.

Here are some more examples:

$$4 \cdot -9 = 1 \cdot -1 \cdot 4 \cdot 9 = -1 \cdot 36 = -36$$

$$-3 \cdot -5 = -1 \cdot -1 \cdot 3 \cdot 5 = 1 \cdot 15 = 15$$

$$-6 \div 2 = \frac{-1}{1} \cdot \frac{6}{2} = -1 \cdot 3 = -3$$

$$4 \div -2 = \frac{1}{-1} \cdot \frac{4}{2} = -1 \cdot 2 = -2$$

$$-9 \div -3 = \frac{-1}{-1} \cdot \frac{9}{3} = 1 \cdot 3 = 3$$

Once you know how this process works, you can skip some of the extra writing involved if you remember that negative signs knock each other out, like this:

$-3 \cdot -5 = --3 \cdot 5 = 15$ (The two negatives cancel.)

$4 \div -2 = +-\frac{4}{2} = -2$ (There's only one negative, so it doesn't cancel.)

—Dr. Math, The Math Forum

ASK DR. MATH FAQ

NEGATIVE × NEGATIVE = POSITIVE

Minus times minus results in a plus,
The reason for this, we needn't discuss.

—Ogden Nash

Why is a negative times a negative a positive?

People have suggested many ways of picturing what is going on when a negative number is multiplied by a negative number. It's not easy to do, however, and there doesn't seem to be a visualization that works for everyone.

If you think of multiplication as repeated addition (which is what everyone is taught), then as long as there is a nonnegative factor in there somewhere, you can visualize doing *something* some number of times (or not at all, if one of the factors is zero)—that is, you can assign a *role* to each of the factors: there's the thing being repeated, which can be positive or negative (going east or west, spending or collecting), and there's the number of repetitions (which has to be positive). No matter where the numbers came from, you can always see the nonnegative factor as the number of repetitions:

$3 \cdot 5$	Go 3 miles east, 5 times.
$-3 \cdot 5$	Go 3 miles west, 5 times.
$3 \cdot -5$	Go 5 miles west, 3 times.
$-3 \cdot -5$	Huh?

When *both* of the factors are negative, you look at both numbers and *neither* of them can be the number of repetitions, because it doesn't make sense to repeat something a negative number of times!

We realize there are folks for whom none of these explanations are going to make sense. It's just a hard concept to get in your head. But for those for whom it will help, here are a few ideas.

Debt

Debt is a good example of a negative number. One common form of debt is a mortgage, in which you owe the bank money because the bank paid for your house. It is common for an employer to deduct a mortgage payment from an employee's paycheck to help the employee keep on schedule with the payments.

The employee's paycheck each month shows a $700 deduction. After 6 months, how much money has been taken out of the paycheck for the mortgage? We can figure out the answer by doing multiplication:

$$6 \cdot -\$700 = -\$4200$$

A positive times a negative results in a negative.

As a bonus, suppose that the employer decides to pay the employee's mortgage for one year. The employer removes the mortgage deduction from the monthly paychecks. How much money is gained by the employee in our example? We can represent "removes" by a negative number and figure out the answer by multiplying:

$$-12 \cdot -\$700 = \$8400$$

A negative times a negative results in a positive. So, the employee gains $8,400 when the employer pays the mortgage.

If you think of multiplication as grouping, then we have made a positive group by taking away a negative number 12 times.

The convention that $(-1)(-1) = +1$ has been adopted for the simple reason that *any other convention would cause other rules to be broken.*

For example, if we adopted the convention that $(-1)(-1) = -1$, the distributive property of multiplication wouldn't work for negative numbers:

$$\begin{aligned}(-1)(1 + -1) &= (-1)(1) + (-1)(-1)\\ (-1)(0) &= -1 + -1\\ 0 &= -2\end{aligned}$$

As Sherlock Holmes observed, "When you have excluded the impossible, whatever remains, however improbable, must be the truth."

Since everything except $+1$ can be excluded as impossible, it follows that, however improbable it seems, $(-1)(-1) = +1$.

A mathematical explanation

Here's a mathematical explanation of why a negative number multiplied by another negative number makes a positive number. If you think of a negative number as just a positive number multiplied by −1, then you can write the product of two negative numbers this way:

$$(-a)(-b) = (-1)(a)(-1)(b) = (-1)(-1)ab$$

Since we know from convention that $(-1)(-1) = +1$ (see above), then $(-a)(-b) = ab$. For example,

$$\begin{aligned}-2 \cdot -3 &= (-1)(2)(-1)(3)\\ &= (-1)(-1)(2)(3)\\ &= (-1)(-1) \cdot 6\end{aligned}$$

Number line

We can use a number line to explain why a negative number multiplied by a negative number equals a positive number. Imagine a number line on which you walk. Multiplying $x \cdot y$ means taking x steps, each of size y. Negative steps require you to face the negative end of the line before you start walking, and negative step sizes are backward (i.e., heel first) steps. So, $-x \cdot -y$ means to stand on zero, face in the negative direction, then take x backward steps, each of size y. This puts you on the positive

Let a and b equal any two real numbers. Consider the number x defined by

$$x = ab + (-a)(b) + (-a)(-b).$$

We can write

$$\begin{aligned} x &= ab + (-a)[(b) + (-b)] \qquad \text{(factor out } -a\text{)} \\ &= ab + (-a)(0) \\ &= ab + 0 \\ &= ab \end{aligned}$$

We can also write that

$$\begin{aligned} x &= [a + (-a)]b + (-a)(-b) \qquad \text{(factor out } b\text{)} \\ &= 0 \cdot b + (-a)(-b) \\ &= 0 + (-a)(-b) \\ &= (-a)(-b) \end{aligned}$$

So, if

$$x = ab$$

and

$$x = (-a)(-b)$$

then by the transitivity of equality (which says that if $a = b$ and $b = c$, then $a = c$, we have

$$ab = (-a)(-b)$$

side of the number line. For example, for (–3)(–2), you would start on zero facing the negative direction and take 3 backward steps of 2 spaces each.

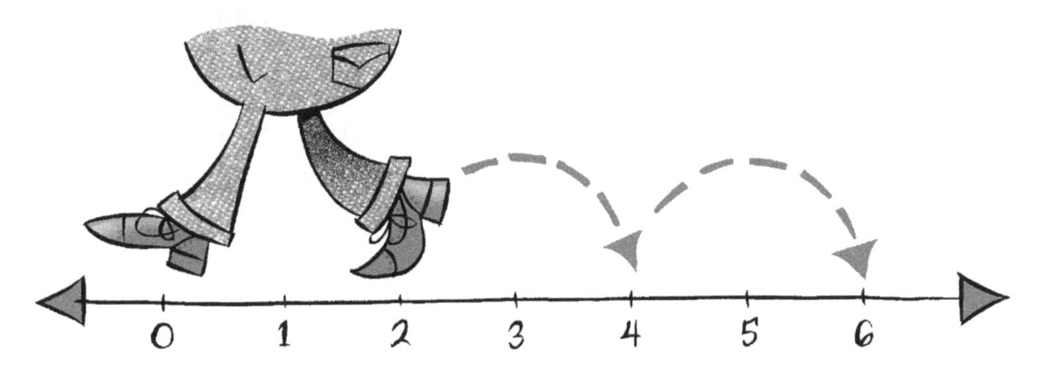

So, (–3)(–2) = 6.

Here's an algebraic mnemonic poem (a mnemonic is something that helps you remember a fact) by Jean Hervé-Bazin, with English text by Julio González Cabillón:

Negative times negative is positive:
The enemies of our enemies are our friends.

Multiplying Two Negative Numbers

Dear Dr. Math,

I know that when you multiply two negative numbers you get a positive answer. I want to know why the answer is positive. Can you give me a written example?

—Carissa

Hi, Carissa,

Suppose you're standing on a road, and you measure mileage to the east as positive and to the west as negative. You are at zero, and a town 1 mile east is at +1 mile, while a town 2 miles to the west is at –2 miles. So far, so good?

A car traveling east will have a positive velocity, and a car traveling west will have a negative one. So, a car going east at 60 mph goes at +60 mph, and a car going west at the same speed goes at –60 mph.

This makes sense, since if they go for an hour (+1 hour), the east-going car will be at (+1)(+60) = 60 miles, and the car going west will be at (+1)(–60) = –60 miles (= 60 miles west). Still okay?

Now, suppose a car passes you going east at 60 mph. Where was it 1 hour ago? Or at –1 hour? Just multiply (–1)(60) = –60 = 60 miles west.

How about a car going west at 60 mph—where was it an hour ago? Its velocity is –60 and the time is –1, so it was at (–1)(–60). The answer is 60 miles east, or +60. So, (–1)(–60) = +60. I hope this helps.

Try making up a similar example with money. Money you earn is positive, and money you spend is negative. If you earn or spend at a constant rate, multiply to find out your change in worth.

—Dr. Math, The Math Forum

Multiplying and Dividing Negatives

Dear Dr. Math,

I need to figure out how to multiply and divide integers. We learned it last year, but I never got it. Please help me.

–Clive

Clive,

First, if the problem contains just multiplication, just division, or just multiplication and division, then the correct sign can be determined rather simply. Just count up how many negative (–) signs there are in the problem. If there is an even number of negative signs, the answer will be positive. If there is an odd number of negative signs, the answer will be negative. Then work the problem as if it had no signs and put the correct sign on the answer.

Example 1: (–2)(–3)(4)(–1). Note that there are three negatives, so the answer is negative. Then $2 \cdot 3 \cdot 4 \cdot 1 = 24$, so (–2)(–3)(4)(–1) = –24.

Example 2: (–5)(7)(–1). Note that there are two negatives, so the answer is positive. Then $5 \cdot 7 \cdot 1 = 35$, so (–5)(7)(–1) = 35.

Example 3: $\frac{(-2)(2)(-4)}{(-1)(-4)(-5)}$. Note that there are five negatives, so the answer is negative. Then $\frac{(2)(2)(4)}{(1)(4)(5)} = \frac{16}{20} = \frac{4}{5}$ or 0.8, so the answer is –0.8.

But remember this method works only if the problem contains just multiplication and division. If it has addition or subtraction in it, you have to figure out what to do with each section of the problem individually.

—*Dr. Math, The Math Forum*

MORE FACTS ABOUT INTEGERS

The terms *even* and *odd* only apply to integers; 2.5 is neither even nor odd. Zero, on the other hand, is even, since it is 2 times some integer: it's 2 times 0. To check whether a number is odd, ask yourself whether it's one more than some even number: 7 is odd, since it's one more than 6, which is even.

Stated another way, zero is even, since it can be written in the form $2 \cdot n$, where n is an integer. Odd numbers can be written in the form $2 \cdot n + 1$. Again, this lets us talk about whether negative numbers are even or odd: –9 is odd, since it's one more than –10, which is even.

Every integer greater than 1 can be factored into the product of prime numbers, and there's only one way to do it for every number. Remember that a prime number is a positive integer that has exactly two positive integer factors: 1 and itself.

For instance, $280 = 2 \cdot 2 \cdot 2 \cdot 5 \cdot 7$, and there's only one way to factor 280 into prime numbers. This is an important theorem called the Fundamental Theorem of Arithmetic. (See Part 3 for more on factoring.)

Most mathematicians, at least when they're talking to one another, use Z to refer to the set of integers. In German, the word *zahlen* means "to count" and *Zahl* means "number." Mathematicians also use the letter *N* to talk about the set of positive integers—that is, the set {1, 2, 3, 4, 5, 6, . . .}.

Absolute Value

Remember that we said the integer –3 consists of the sign "–" and the absolute value 3? Absolute value is a way of saying ignore the sign. This section will show you how to work with absolute value in different situations.

Evaluating Absolute Values

Dear Dr. Math,

How do you answer a question that looks like $|-6 + 5|$?

—Clive

Hi, Clive,

When an expression is surrounded by two vertical bars, as in your case, that means take the absolute value of the expression. The absolute value of a number is the distance of that number from zero. For example, $|2| = 2$ and $|-3| = 3$. A mathematical definition of absolute value is

$$\begin{aligned} |x| &= x \text{ if } x \geq 0 \\ &= -x \text{ if } x < 0 \end{aligned}$$

If this confuses you, don't worry. An easy way to take the absolute value of an expression is to start by evaluating the expression (add, subtract, multiply, divide, or complete any other noted operations), then drop the minus sign if it's negative. If it's positive, do nothing. For example,

Step 1: $|-4 + 2| = |-2|$	In this step, evaluate the expression.
Step 2: $|-2| = 2$	In this step, take the absolute value. Since –2 is negative, drop the minus sign.

Here's another example:

Step 1: $\lvert -4 - (-5) \rvert = \lvert -4 + 5 \rvert = \lvert 1 \rvert$	In this step, evaluate the expression.
Step 2: $\lvert 1 \rvert = 1$	In this step, take the absolute value. Since 1 is positive, $\lvert 1 \rvert = 1$.

It's important that you always remember to evaluate the expression *before* dropping any minus signs.

Now, it should be easy for you to evaluate $\lvert -6 + 5 \rvert$.

—Dr. Math, The Math Forum

Negative Absolute Value

Dear Dr. Math,

My friend and I are wondering if $-|8|$ is -8 or 8. We have different opinions on the answer.

Thanks,

Carissa

Dear Carissa,

$-|8| = -8$. The minus sign acts after the absolute value is taken. $|8|$ is 8. Then put a minus sign in front of the 8 and you get -8. Note that $-|-8|$ is also -8. This is because the absolute value of -8 is 8, then the minus sign goes out front.

Absolute value bars are sort of like parentheses or vinculums (which are the horizontal lines dividing the **numerators** and **denominators** of fractions)—they're grouping symbols. The general rule is that whenever something is grouped, you have to figure out its value before you can apply operators, even negative signs, to it.

—Dr. Math, The Math Forum

PRACTICAL APPLICATIONS OF ABSOLUTE VALUE

Here are a few real-world examples of the concept of absolute value:

1. Distances in real life: suppose you go 3 blocks east, then 6 blocks west, then 11 blocks east again. We can ask two questions: (1) Where are you relative to where you started? This requires us to retain the sign information and is not answered by the absolute value. (2) How far did you travel? To find the answer, add $3 + 6 + 11$, because despite the fact that you've retraced your steps a couple of times, you have traveled 20 blocks altogether.

   Of course, distances are useful in many other real-world applications, such as navigation and transport (do we have enough fuel to get there and back?), architecture, engineering, science, and sports. If a football team starts with the ball at midfield, how many consecutive 15-yard penalties can the referees call before the next penalty would be half the distance to the goal line?

2. Suppose you are driving a car. Going too fast is obviously a hazard and you might get a speeding ticket. Going too slow is also a hazard and can get you a ticket. What matters most is how different your speed is from everyone else's speed. This type of difference is fundamental to all sorts of concepts in statistics, where the absolute value is used in various ways of quantifying how well or how poorly one thing predicts another. Statistics is used in many important real-world applications including medicine and finance.

3. Suppose you are exchanging currency, say dollars and pesos. The bank or exchange will charge a commission based on how much is exchanged (sometimes there will be a flat fee as well). This commission is applied whether you buy dollars or pesos. In your pocket you may have D dollars and P pesos, and D could go up (or down) while P goes down (or up). But the commission on a given transaction is the absolute value of either D or P, times the exchange rate.

Resources on the Web

Learn more about integers at these sites:

Math Forum: Graphing Linear Functions

mathforum.org/alejandre/linear.graph.html

Directions to make graphs of linear functions using a Claris Works spreadsheet.

Math Forum: Operations with Integers

mathforum.org/alejandre/frisbie/student.math.stars.html

Use the shareware program Math Stars, by Roger Clary, to practice addition, subtraction, multiplication, and division of integers.

Math Forum: Problem of the Week: An Interesting Integer

mathforum.org/midpow/solutions/solution.ehtml?puzzle=83

I'm thinking of a particular integer. It is a multiple of 3, 5, and 7. No digit occurs more than once. Can you find my number?

Math Forum: Problem of the Week: Temperature Change

mathforum.org/midpow/solutions/solution.ehtml?puzzle=20

Use the given information to find the temperature.

Shodor Organization: Project Interactivate: Cartesian Coordinate System

shodor.org/interactivate/lessons/cartesian.html

Students become familiar with the Cartesian Coordinate System and its many uses in the world of mathematics.

Shodor Organization: Project Interactivate: Graphing and the Coordinate

shodor.org/interactivate/lessons/fm1a.html

A lesson designed to introduce students to graphing ordered pairs of numbers on the coordinate plane.

Shodor Organization: Project Interactivate: Graphs and Functions

shodor.org/interactivate/lessons/fm3.html

This lesson is designed to introduce students to graphing functions.

Real Numbers

Once we have a set of numbers and operations, it's natural to start looking at what happens when we apply the operations to the numbers. In some cases, we just get more of the same types of numbers. For example, by adding or subtracting or multiplying two integers, we can only get another integer ($3 + 5 = 8$, $195 - 201 = -6$, $-4 \cdot 21 = -84$). Mathematicians say that the set of integers is *closed* under these three operations, because the answer to any problems using those operations is another integer. But division is a little different. Although some divisions of integers by integers result in other integers, most do not.

So, what do we do about this? Recall how we *define* division. When we say that $\frac{15}{6}$ = something, there is *some* number such that $6 \cdot$ something $= 15$. One way that we can handle the situation is to just leave the division undone and use the implied division itself as a new type of number. Because each of these new numbers is a **ratio** of two integers, we call them **rational numbers.**

We know where to find integers on a number line. And rational numbers that aren't also integers are located in between the integers. For example, $\frac{15}{6}$ has to be somewhere to the right of 2 (since $2 \cdot 6$ is 12) and to the left of 3 (since $3 \cdot 6$ is 18).

Can we find a rational number at *every* point on the number line? (Which is to say, are *all* numbers rational?) No; lots of numbers can't be represented as a ratio of two integers. You've probably already seen a few of these, like pi, and the square root of 2.

So, we can divide numbers into two classes: rational and not rational. **Irrational numbers** are numbers that aren't rational. The rational and irrational numbers together make up the **real numbers.**

In this part, Dr. Math explains

- Rational numbers
- Factoring
- Squares and square roots
- Irrational numbers
- Pi
- Real numbers

1 Rational Numbers

If a rational number is any number that can be written as a ratio of two integers, then an irrational number is any number that cannot be written this way. This section will help you become more familiar with rational numbers.

Repeating Decimals

```
Dear Dr. Math,
Is .999 . . . a rational number?
Sincerely,
Carissa
```

Dear Carissa,

A number is rational if it can be written as $\frac{a}{b}$, where a and b are integers. For example, $.3 = \frac{3}{10}$. With repeating decimals, you need to figure out how to change them into the form $\frac{a}{b}$ (see the Ask Dr. Math FAQ below). Once you do, you'll see that $.\overline{9}$ (.999 . . . or .9 repeating) is equal to $\frac{9}{9}$, which is equal to 1. We have just written down $.\overline{9}$ in the form $\frac{a}{b}$ where a and b are both 9, which is an integer, so $.\overline{9}$ is a rational number. In fact, all repeating decimals are rational.

—*Dr. Math, The Math Forum*

Ask Dr. Math FAQ: WHY DOES 0.999 . . . = 1 ?

There's no doubt that this equality is one of the weirder things in mathematics. No matter how many 9's you add, you'll never get all the way to 1. Right? But what if you think about moving *away* from 1? If you start at 1 and try to move away from 1 toward .9, how far do you have to go to get to $.\overline{9}$? Any step you try to take will be too far, so you can't really move at all—which means that to move from 1 to $.\overline{9}$, you have to stay at 1!

Here's another way to think about it. When you write something like 0.35, that's the same as $\frac{35}{100}$:

$$0.35 = \frac{35}{100}$$

You can turn that into a repeating decimal by dividing by 99 instead of 100.

$$0.35353535\ldots = 0.\overline{35} = \frac{35}{99}$$

In general, when we have n repeating digits, the corresponding fraction is

$$\frac{\text{(the digits)}}{10^n - 1}$$

Again, some examples can help make this clear:

$$0.\overline{1} = \frac{1}{9}$$

$$0.\overline{12} = \frac{12}{99}$$

$$0.\overline{123} = \frac{123}{999}$$

and so on.

So, here's something to consider: What fraction corresponds to

$$0.\overline{9} = ?$$

It has to be something over 9, right?

$$0.\overline{9} = \frac{?}{9}$$

The *only* thing it could possibly be is

$$0.\overline{9} = \frac{9}{9}$$

But that's the same as 1. Ultimately, though, this probably won't really make sense until you come to grips with what it means for a decimal to repeat *forever,* instead of just for a really long time.

When you think of 0.999 . . . as being a little less than 1, it's because in your mind you've stopped expanding it—that is, instead of

0.999999 . . .

you're really thinking of

0.999 . . . 999

which is not the same thing. You're absolutely right that 0.999 . . . 999 is a little less than 1, but 0.999999 . . . doesn't fall short of 1 *until* you stop expanding it. But you never stop expanding it, so it never falls short of 1.

Suppose someone gives you $1,000 but says, "Now, don't spend it all, because I'm going to go off and find the largest integer, and after I find it I'm going to want you to give me $1 back." How much money has he really given you?

On the one hand, you might say, "He's given me $999, because he's going to come back later and get $1." On the other hand, you might say, "He's given me $1,000, because he's *never* going to come back!"

It's only when you realize in this instance *later* is the same as *never* that you can see that you get to keep the whole $1,000. In the same way, it's only when you really understand that the expansion of 0.999999 . . . *never* ends that you realize it's not really a little less than 1 at all.

Factoring

You might be familiar with factoring from multiplication. Whenever you can multiply two integers to get another integer, the first two are **factors** of the third. For example, $4 \cdot 3 = 12$, so 3 and 4 are factors of 12. However, they're not its only factors; 1, 2, 6, and 12 are other factors of 12. Another way of defining a factor is a number that evenly divides the number you're factoring. For this reason, factors are also referred to as *divisors*. For example, 6 is a factor (or divisor) of 18, because $\frac{18}{6} = 3$. However, 7 is not a factor of 18, because dividing 18 by 7 leaves a remainder.

Prime Factorization

```
Dear Dr. Math,
Could you give me a crash course on prime
factorization? We just started that section
at school today, and I know nothing about it.
  In class, some kids can divide so quickly—
is there a trick to finding factors
quickly?
—Clive
```

Dear Clive,

Prime factorization is pretty simple once you know what's going on, and it can be fun, too! I used to do it in the margins of my notebooks when I was bored in class.

Every whole number can be put into one of two categories: prime or composite. **Prime numbers** are numbers that you can't divide without getting a fraction (unless you divide them by themselves or by 1). These are numbers like 3, 5, 7, 19, or 379,721 (I told you I used to do this when I was bored!). **Composite numbers,** on the other hand, can be split up into some smaller whole numbers: 8 is 2 times 4, 4 is 2 times 2, 21 is 3 times 7, 100 is 5 times 5 times 2 times 2, and so on. It's easy to remember which word means which—composite has the word *compose* in it, and you can make composite numbers out of smaller numbers just like you can compose a piece of music out of smaller groups of notes.

(Note that the definition of a prime number doesn't allow 1 to be a prime number: 1 has only one factor, namely 1. Prime numbers have *exactly* two factors.)

Prime factorization simply takes a composite number and splits it up into smaller numbers as much as possible—that is, into primes. Take the number 8, for example. We can divide 8 into 2 and 4.

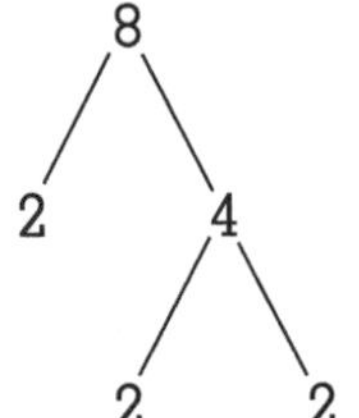

But we're not done yet; 4 can be split into 2 and 2 . . .

Now we're done because the 2's can't be split anymore.

If you draw the smaller pieces (called the factors) in an upside-down tree like I just did, you can go back and collect all the pieces at the ends of the branches; they will be the prime factors of the number. So, the prime factors of 8 are 2, 2, and 2.

Here's another example:

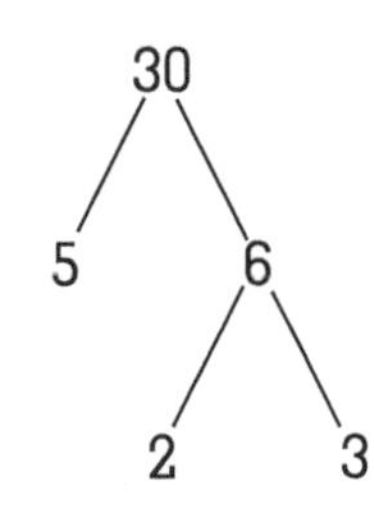

So, the prime factors of 30 are 2, 3, and 5.

Now, you wanted to know some tricks for factoring:

2—Every even number is divisible by 2; if a number ends in a 2, 4, 6, 8, or 0, at least one of its factors will be a 2.

3—If you add the digits of a number and the number you get is divisible by 3, then the original number is divisible by 3. For example, take the number 57. You add the digits, $5 + 7 = 12$. Since 12 is divisible by 3, 57 is divisible by 3. So, 3 is a factor of 57.

4—Take the last two digits of a number. If they are divisible by 4, then the number is divisible by 4. For example, 216 is divisible by 4 because 16 is divisible by 4. Since 4 is $2 \cdot 2$, you know that two 2's are prime factors of this number.

5—Any number that ends in a 5 or a 0 is divisible by 5.

6—If the number is divisible by both 3 and 2, it is divisible by 6 as well.

7—To find out if a number is divisible by 7, take the last digit, double it, and subtract it from the rest of the number. (For example, if you had 203, you would double the last digit to get 6, and subtract that from 20 to get 14.) If you get an answer divisible by 7 (including 0), then the original number is divisible by 7. If you don't know the new number's divisibility, you can apply the rule again.

8—If the last three digits of the number are divisible by 8, the number is divisible by 8. Since 8 can be factored into $2 \cdot 2 \cdot 2$, you know that three 2's will be prime factors of the number.

9—If the sum of the digits of the number is divisible by 9, then the number is divisible by 9. Since $3 \cdot 3 = 9$, you know that two 3's are prime factors of the number.

10—If a number ends with a 0, it's divisible by 10. Since 10 is $5 \cdot 2$, you know 5 and 2 will be prime factors of that number.

There are plenty more shortcuts out there (maybe you could try your hand at finding a few!), but that should be enough to get you started.

—*Dr. Math, The Math Forum*

A number is the *product* of its *factors:* $24 = 4 \cdot 6$, so 4 and 6 are factors of 24. Also $24 = 3 \cdot 8$, so 3 and 8 are factors of 24.

When we talk about factors, we're only talking about integers, and usually only positive integers. So, although it's true that $18 = 4 \cdot 4.5$, we wouldn't say that 4 and 4.5 are factors of 18. Also, it's true that $18 = -3 \cdot -6$, but we wouldn't normally say that -3 and -6 are factors of 18.

Often the words *factor* and *divisor* are used as synonyms, but some people would say that factors can only come in groups, whereas a divisor can stand alone. These people would say that 3 and 6 are factors of 18; but 3 is a divisor of 18, and 6 is a divisor of 18. They may roll their eyes if you fail to make this distinction, but they'll understand what you mean.

Some people assume that *factor* means *prime factor*—they'll tell you the factors of 12 are 2 and 3 (or 2 and 2 and 3), instead of 1, 2, 3, 4, 6, and 12. It's wise to ask if someone gives you a list that seems short to you. *Proper factors* are all the factors of a number except the number itself. The proper factors of 12 are 1, 2, 3, 4, and 6.

A *factorization* of a number is any expression of the number as the product of a set of factors. For example, some factorizations of 24 are

$$24 = 1 \cdot 24$$
$$24 = 2 \cdot 2 \cdot 6$$
$$24 = 2 \cdot 3 \cdot 4$$
$$24 = 2 \cdot 2 \cdot 2 \cdot 3$$

A factorization in which each factor is a prime number is called a *prime factorization.* Each number has exactly one prime factorization unique to that number. Later on, we'll talk more about prime numbers and prime factorizations and why they're important.

Writing Numbers as Products of Prime Factors

Dear Dr. Math,

I need to write 4, 6, 8, 10, 12, and 14 as products of their prime factors. Please help! My mind is a blank.

—Clive

Dear Clive,

Recall that prime numbers are divisible only by themselves and 1. Examples are 2, 3, 5, 7, 11, and 13.

The prime factors of a number are prime numbers such that their product is the given number. So, you need to do two things:

1. Find the prime numbers that are factors of the given number.
2. Write the given number as a product of its prime factors.

To obtain the prime factors of a number, first divide the number by any of the prime numbers and see which divide the given number with zero as a remainder. Begin by using the smaller prime numbers because they are easier. When you find a prime number that divides your number, you get a certain quotient. You then need to find a prime factor of the quotient by the same process.

For example, let's try this with number 12. We begin with the smallest prime number: 2. We can divide 2 into 12 to get 6 as a quotient and zero as a remainder, so we can say that $12 = 2 \cdot 6$.

We have now found two factors of 12—2 and 6—but they are not both prime factors. Therefore, we now need to write 6 as a product of prime factors.

Again we begin with the smallest prime number: 2. We divide 6 by 2 to get 3 as a quotient and zero as a remainder, so we can say that $6 = 2 \cdot 3$. The divisions end here because the quotient obtained, 3, is a prime number.

Now, how do we write 12 as a product of its prime factors (2 and 3 in this case)? We could just write $12 = 2 \cdot 2 \cdot 3$, but a shorter way would be $12 = 2^2 \cdot 3$, where 2^2 means 2 squared or 2 raised to the second power.

Another method to find the prime factors is to use factor trees. Find any factors of your number, then factors of these factors, and so on, until you have only prime numbers.

For example, if you want to factor 24, you may know that $24 = 6 \cdot 4$. You easily factor 6 and 4 to get $6 = 2 \cdot 3$ and $4 = 2 \cdot 2$. These factors are all prime numbers so you can stop here. The factor tree looks like this:

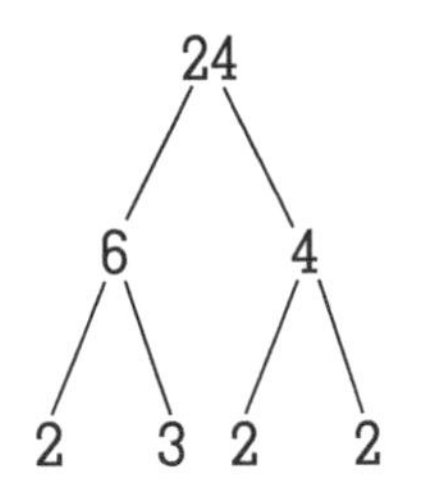

Since the prime factor 2 shows up 3 times, you can write this as 2^3, so your answer is

$$24 = 2^3 \cdot 3$$

—*Dr. Math, The Math Forum*

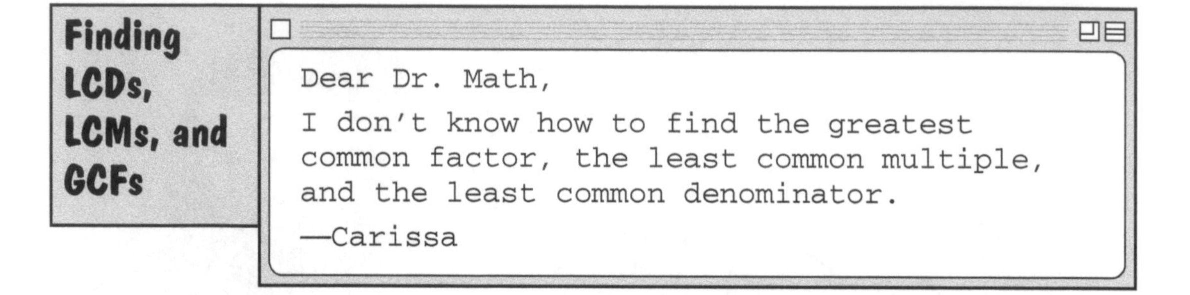

Dear Carissa,

Let's start with the **greatest common factor (GCF).** The GCF of two (or more) numbers is the product of all the factors that the numbers have in common. For example, to find the greatest common factor of 32 and 76, you would first express both as products of their prime factors, then look for factors common to both:

$$32 = 2 \cdot 2 \cdot 2 \cdot 2 \cdot 2$$
$$76 = 2 \cdot 2 \cdot 19$$

There are two 2's common to both numbers, so $2 \cdot 2 = 4$ is the greatest common factor of 32 and 76.

Now let's find the **least common multiple (LCM).** The LCM of two (or more) numbers is the product of one number times the factors of the other number(s) that aren't common to both. If you want to find the least common multiple of 32 and 76, first find the prime factors, as in the example above. Then choose either number, and look at the factors of the other number. I'll pick 32. There's only one factor of 76 that does not appear in the factorization of 32: 19. So, multiply 32 by 19, and the result is the LCM. If you choose 76 as the number to multiply, look at the factors of 32. They're all 2's—but only two of them are common to the factorization of 76. There are three more than that in the factorization of 32. Multiply 76 by those three 2's, and the result is the LCM.

Finally, let's look at the **least common denominator (LCD).** The LCD of two or more fractions is the smallest number that can be divided into each denominator with no remainder. The least common denominator can also be thought of as the least common multiple of two (or more) different denominators. If you know how to find least common multiples, you can find least common denominators. Once you've found a least common multiple for the denominators, you reexpress each fraction so that its denominator is the least common multiple by multiplying by a fraction equal to 1 (remember that multiplying by 1 doesn't change the value of a number).

Rx **When you factor, it's always a very good idea to list the factors in increasing order, because it makes it much easier to see *common factors*—the factors that two or more numbers have in common.**

For example, in order to add $\frac{1}{6}$ and $\frac{1}{8}$, you first need to find the least common denominator. To do this, first find the least common multiple of the denominators 6 and 8. Since $6 = 2 \cdot 3$ and $8 = 2 \cdot 2 \cdot 2$, the least common multiple is $8 \cdot 3$ or 24. Then multiply $\frac{1}{6}$ by $\frac{4}{4}$ and multiply $\frac{1}{8}$ by $\frac{3}{3}$ to reexpress both $\frac{1}{6}$ and $\frac{1}{8}$ as some number of 24ths:

$$\frac{1}{6} \cdot \frac{4}{4} = \frac{4}{24}$$

$$\frac{1}{8} \cdot \frac{3}{3} = \frac{3}{24}$$

Now you can add the fractions

$$\frac{4}{24} + \frac{3}{24} = \frac{4+3}{24} = \frac{7}{24}$$

So,

$$\frac{1}{6} + \frac{1}{8} = \frac{7}{24}$$

Remember, whether you're trying to find the greatest common factor or the least common multiple, the key is factoring the numbers to their prime factors, then noting which factor(s) the numbers have in common.

The product of the common factors will be the greatest common factor. The product of one number and the factors of the other that are *not* common to the first will be the least common multiple.

—Dr. Math, The Math Forum

Unknowns

```
Dear Dr. Math,
Could you explain the concept of factoring
unknowns?
Thank you for your help,
Clive
```

Dear Clive,

The sort of factoring you are doing is somewhat similar to regular factoring, but since it's algebra, there are letters like x and y stuck into the equations. These letters just stand for unknowns.

For example, you could have an equation like $x + 1 = 5$. You could ask, "What number added to 1 gives you 5?" That number is represented by x.

You could figure out the equation $x + 1 = 5$ by subtracting 1 from each side:

$$\begin{array}{r} x + 1 = 5 \\ -1 \;\; -1 \\ \hline x = 4 \end{array}$$

Other algebraic equations (equations with letters) might look different, but the idea behind them is the same.

So, what does this have to do with factoring? Sometimes you'll have an equation that has an unknown squared or cubed term. Here's an example: $x^2 = 9$. Since the x is squared, there are two possible answers for x. You can probably guess that the answers are –3 and 3, since $-3 \cdot -3 = 9$ and $3 \cdot 3 = 9$.

But suppose you don't know what the answer is (or suppose that you're dealing with a more complicated equation). How would you figure out what the answer is? Here's where we get back to factoring. This is what you'd do:

$$\begin{array}{r} x^2 = 9 \\ -9 \quad -9 \\ \hline x^2 - 9 = 0 \end{array}$$

Here's the trick. Remember, if you multiply 5 by 0, you get zero. In fact, if you multiply 28, or 1 billion, or any other number by 0, you get zero. And if you multiply and get zero as an answer, at least one of the numbers you multiplied by *must have been zero!*

We can use that fact here. We know that $x^2 - 9 = 0$. So, one of the factors of $x^2 - 9$ must equal zero.

Next we factor. Basically, this means asking, "What do we multiply to get $x^2 - 9$?" The factors of $x^2 - 9$ are $x - 3$ and $x + 3$, so $(x - 3)(x + 3) = 0$. That means either $x - 3 = 0$ or $x + 3 = 0$.

If $x - 3 = 0$, then

$$\begin{array}{r} +3 \quad +3 \\ \hline x = 3 \end{array}$$

If $x + 3 = 0$, then

$$\begin{array}{r} -3 \quad -3 \\ \hline x = -3 \end{array}$$

So, the answer is, *x* is 3 or –3.

—**Dr. Math, The Math Forum**

Squares and Square Roots

The idea of **square roots** originally comes from geometry: if you have a square with area 9, how long is each of its sides? Since the area of a square is side times side, the sides of this square are each 3 units long. So, 3 is called the square root of 9. The notation that we use is $\sqrt{9} = 3$. It follows from this idea that we write $3 \cdot 3$ as 3^2 or "three squared" and $3^2 = 9$.

What Are Square Roots?

Dear Dr. Math,

What are square roots and how do they work?

Sincerely,

Clive

Dear Clive,

You asked a great question. I'm very impressed that you are already thinking about square roots—you are indeed a budding mathematician.

Before I answer your question about square roots, I think it would be best to talk about square numbers. A *square number* is what you get when you multiply a whole number by itself. For instance, since $3 \cdot 3 = 9$, 9 is a square number. Likewise, 16 is a square number, since $4 \cdot 4 = 16$. Can you think of some other square numbers?

Taking the square root of a number is the opposite operation (in math lingo it is called the **inverse**) of squaring a number. To find the square root of a number, you ask yourself what number times itself will yield that number. So, say you wanted to know what the square root of 9 was. You would ask yourself what number multiplied by itself yields 9. Then you would calculate that $3 \cdot 3 = 9$, so 3 is the square root of 9. And you would be right.

But there is also another square root of 9. It's –3, because $-3 \cdot -3 = 9$. One of the most basic facts about real numbers is that a negative number times a negative number yields a positive number. That is why $-3 \cdot -3 = 9$.

But although every positive number has two square roots, we usually only worry about the positive square root. So, if someone were to ask you what the square root of 9 is, you could just say 3 (and not worry about –3).

If you progress another step and think about the square roots of numbers that are not square, then things get a little messy; but not to fear, the same ideas will still work. When we try and find the square root of a number that is not square, we can either **approximate** it using a calculator (or by doing calculations by hand using trial and error) or use the square root sign and leave the square root sign in the reduced form of the number. Say, for instance, you were wondering how to write the square root of 3. The symbol is 3 under the square root sign: $\sqrt{3}$, which indicates that when that number, $\sqrt{3}$, is squared, it yields 3.

Rx **You can picture square numbers by looking at a square on a grid with sides of the unit that is being squared. If the square has sides of 6 units, how many units make up the square's area? Try it and count the units.**

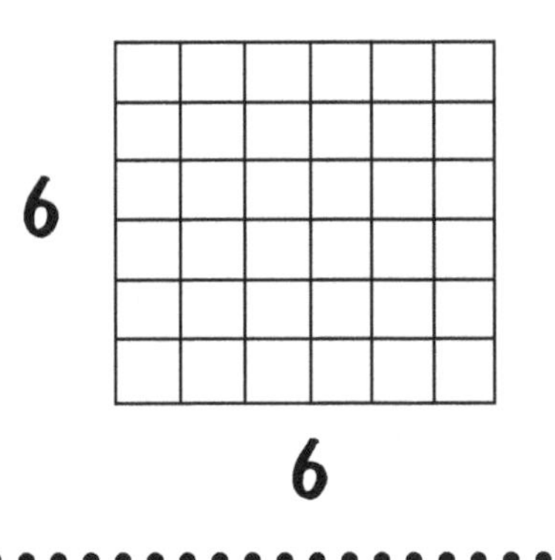

—Dr. Math, The Math Forum

ASK DR. MATH FAQ: SQUARE ROOTS WITHOUT A CALCULATOR

How do you figure out square roots without using a calculator?

Start with the number you want to find the square root of. Let's use 12. There are three steps:

1. Guess
2. Divide
3. Average

. . . and then just keep repeating steps 2 and 3.

For step 1, start by guessing a square root value. It helps if your guess is a good one, but it will work even with a terrible guess. We will guess that 2 is the square root of 12.

For step 2, we divide 12 by our guess of 2 and get 6.

For step 3, we average 6 and 2:

$$\frac{6 + 2}{2} = 4$$

We repeat step 2 by dividing 12 by the new guess of 4. So,

$$\frac{12}{4} = 3$$

Now average 4 and 3:

$$\frac{4 + 3}{2} = 3.5$$

Divide 12 by the new guess:

$$\frac{12}{3.5} = 3.43$$

Now take the average:

$$\frac{3.5 + 3.43}{2} = 3.465$$

We could keep going forever, getting a better and better approximation, but let's stop here to see how we are doing:

$$3.465 \cdot 3.465 = 12.006225$$

That is quite close to 12, so we are doing pretty well.

(See mathforum.org/alejandre/escot/numbers.html)

4 Irrational Numbers

Irrational numbers are numbers that can be approximated with decimals but not written as fractions using integers. The square root of 2 is an irrational number because it can't be written as a ratio of two integers. Other irrational numbers include the square root of 3, the square root of 5, pi, *e*, and the golden ratio.

Pi (π) is an irrational number because it cannot be precisely expressed as a ratio of two integers: it has no *exact* decimal equivalent, although 3.1415926 is good enough for almost any application, while 3.14 and $\frac{22}{7}$ are good enough for when you just need an **estimate.**

The golden ratio is a special number approximately equal to 1.6180339887498948482. . . . It is an irrational number with an unending series of digits, so it is often better to use its exact value given by

$$\frac{1 + \sqrt{5}}{2}$$

A golden rectangle is a rectangle in which the ratio of the length to the width is the golden ratio. In other words, if one side of a golden rectangle is 2 feet long, the other side will be approximately equal to $2 \cdot (1.62) = 3.24$ feet.

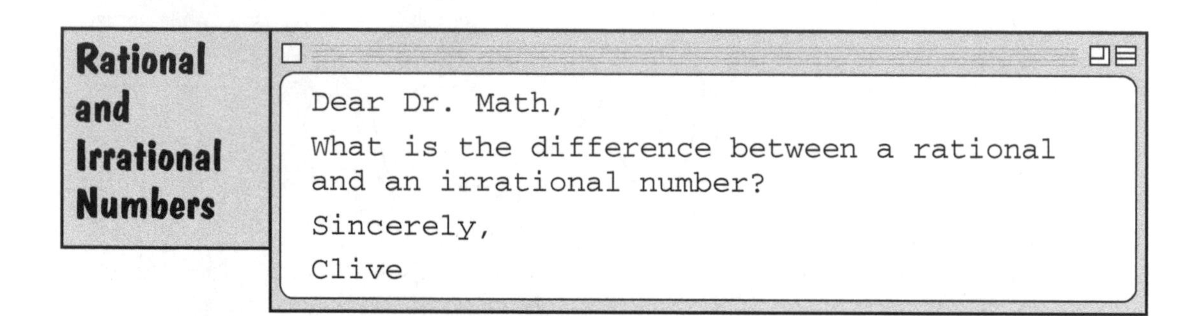

Rational and Irrational Numbers

Dear Dr. Math,

What is the difference between a rational and an irrational number?

Sincerely,

Clive

Dear Clive,

A rational number is any number that can be written as a ratio of two integers—that is, a number is rational if we can write it as a fraction in which the numerator and denominator are both integers. Every integer is a rational number, since each integer n can be written in the form $\frac{n}{1}$. For example, $3 = \frac{3}{1}$, so 3 is a rational number. However, numbers like $\frac{1}{4}$, 358,903/75,980, and $-\frac{8}{9}$ are also rational, since they are fractions in which the numerator and denominator are integers. Numbers like $\frac{180}{\pi}$ and $\sqrt{2}/2$ are fractions, but they are not rational, since π and $\sqrt{2}$ are not integers.

An irrational number is any real number that is not rational. By "real" number, I mean, loosely, a number that we can conceive of in this world—one with no square roots of negative numbers (those are called *complex numbers*—you'll get to them in later math classes). A real number is a number that is somewhere on the number line.

So, any number on the number line that isn't a rational number is irrational. For example, the square root of 2 is an irrational number because it is a real number that can't be written as a ratio of two integers. Other irrational numbers include the square root of 3, the square root of 5, and pi.

—Dr. Math, The Math Forum

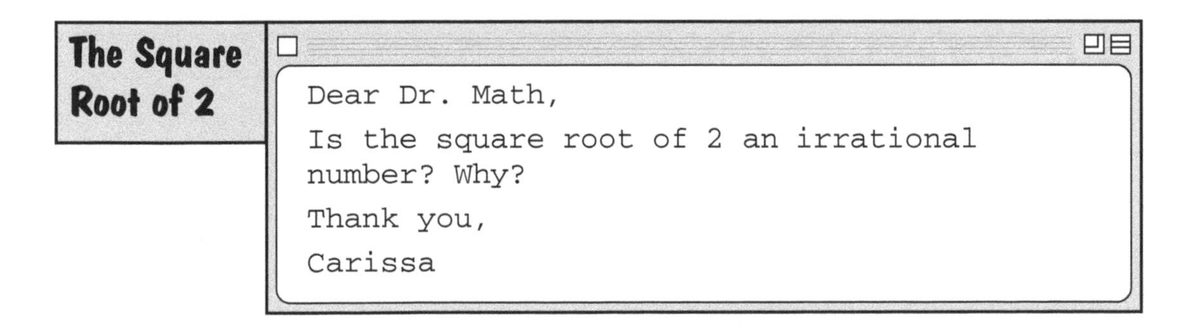

The Square Root of 2

Dear Dr. Math,

Is the square root of 2 an irrational number? Why?

Thank you,

Carissa

Dear Carissa,

A rational number is one that can be expressed as $\frac{a}{b}$ where a and b are integers. The square root of 2 cannot be expressed as $\frac{a}{b}$, so it is an irrational number. Here's the way the proof is usually written: Start with two numbers a and b, such that they have no common factors (i.e., $\frac{a}{b}$ is in lowest terms), so $\frac{a}{b} = \sqrt{2}$.

If $\frac{a}{b} = \sqrt{2}$, then $a^2/b^2 = 2$, which means that a^2 must be even. So, a must also be even and a^2 must be divisible by 4, which means that b^2 must be even as well. So, b must be even. But then $\frac{a}{b}$ must not be a fraction in its lowest terms because a and b are even numbers, and are therefore divisible by 2. Since I asked you for a fraction in lowest terms, that's a contradiction.

The proof finds a hidden contradiction in the assumption that the square root of 2 can be written as a fraction $\frac{a}{b}$ in lowest terms.

—Dr. Math, The Math Forum

5 Pi

By definition, pi is the ratio of the **circumference** of a circle to its **diameter**—that is, the ratio of the distance around to the distance across the circle. Pi is always the same number, no matter which circle you use to compute it. For the sake of usefulness, people often need to approximate pi. For many purposes, you can use 3.14159, which is really pretty good; but if you want a better approximation, you can use a computer to get it. Here's pi to many more digits: 3.14159265358979323846.

Facts about Pi

```
Dear Dr. Math,
What are a few interesting facts about pi?
Thank you,
Carissa
```

Dear Carissa,

Pi (π) was known by the Egyptians, who calculated it to be approximately $(\frac{4}{3})^4$, which equals 3.1604. The earliest known reference to pi occurs in a Middle Kingdom papyrus scroll written around 1650 B.C. by a scribe named Ahmes. Toward the end of the scroll, which is composed of various mathematical problems and their solutions, the area of a circle is found using a rough sort of pi.

Around 200 B.C., Archimedes of Syracuse, a Greek mathematician, found that pi was about 3.14 (in fractions; the ancient Greeks did not have decimals) using the formula $A = \pi r^2$, where A is the area of a circle and r is the **radius** of that circle. Archimedes wrote a book called *The Measurement of a Circle* in which he stated that pi is a number between $3\frac{10}{71}$ and $3\frac{1}{7}$. He figured this out by taking a polygon with 96 sides and inscribing a circle inside the polygon.

Rx **In 1767, a mathematician named Johann Heinrich Lambert proved that pi was an irrational number. Irrational numbers are numbers that do not terminate or repeat when written out as decimals.**

After Archimedes' discovery, there were no further developments or discoveries about pi until the seventeenth century. Pi was called the Ludolphian number, after Ludolph van Ceulen, a German mathematician. The first person to use the Greek letter π for the number was William Jones, an English mathematician, who coined it in 1706.

In the 1800s, people worked for years on end to find the value of pi to about 1,000 places. Imagine doing this by hand with no calculators!

In the twentieth century, there were two important developments: the invention of electronic computers and the discovery of much more powerful formulas for pi. For example, in 1910 the great Indian mathematician Ramanujan discovered a formula that in 1985 was used to compute pi to 17 million digits.

Rx **Mathematicians chose pi as the letter to represent the number 3.141592 . . . , rather than some other Greek letter like alpha or omega because it's pi as in perimeter—the letter pi (π) in Greek is like our letter *p*.**

Recently, even better methods have been developed, and computers are getting ever more powerful. The current record is about 51 billion decimal places. But all the digits of pi can never be known.

—*Dr. Math, The Math Forum*

Real Numbers

To understand what is meant by *real numbers,* it helps to remember that you already know a lot about the different sets of numbers that make up the reals. Every counting number is also a whole number; every whole number is also an integer; and every integer is also a rational number. So, if you combine all that you know about those numbers and include irrational numbers, you already know a lot about the set of real numbers.

Types of Real Numbers

```
Dear Dr. Math,
I would like to know all the types of real
numbers.
Sincerely,
Clive
```

Dear Clive,

There are many, many types of real numbers. I'll tell you about a few here.

The most general division of the real numbers is between the rational and the irrational numbers. A rational number is any number that can be expressed as $\frac{p}{q}$, where p and q are both integers (we'll come to them in a second) and q does not equal zero. You can recognize a rational number because it either has only a finite number of decimal places (or no decimal places at all) or the decimal places repeat themselves. For example, the following are all rational numbers:

348100
78.33954001
3.33333333 . . .
.000012323232323 . . .

But you can make some numbers that don't follow those rules—that is, you could have a number like 1.0100100010000100000 1 . . . which goes on forever but never repeats itself. These are called irrational numbers, and most aren't that pretty. Two of the more famous ones are

$\pi = 3.14159\ldots$
$e = 2.718\ldots$

You can further subdivide the rationals. Here are some ways:

Even numbers and odd numbers

Negative numbers, zero, and positive numbers

Integers and nonintegers

Roughly speaking, integers are numbers with no decimal points, like 4 or 10 or –3788882 or 0. I say roughly speaking, because $\frac{9}{9}$ is an integer, and so is 3.00. These numbers are integers because they are equal to something without a decimal point. Nonintegers are everything else (irrational numbers are also nonintegers).

Then there are whole numbers and natural numbers (also called counting numbers). Natural numbers are positive integers, like 1, 2, 3, 4. . . . Whole numbers are natural numbers and zero.

There are also numbers called *imaginary numbers,* which are outside of the real numbers. These come about when you ask questions like, "What is the square root of –4?" The fundamental unit of the imaginary numbers is *i,* which is defined to be the number such that $i^2 = -1$. Thus, the square root of –4 is $2i$ (can you see why?). Imaginary numbers are a very interesting subject, which if you continue in math, you'll learn about soon enough.

In a branch of math called number theory, you can learn a lot about various types of numbers, such as prime numbers, perfect numbers, square numbers, triangular numbers, and so on.

—Dr. Math, The Math Forum

The Real Number System in a Venn Diagram

Dear Dr. Math,

My teacher gave us an assignment to construct a Venn diagram that illustrates the real number system. I have no idea where to begin and was hoping you could at least identify all the parts of the real number system for me and how they relate to one another so that I can have some point of reference.

Sincerely,

Carissa

Dear Carissa,

I'll give you an explanation of the real numbers first: Real numbers are in a set made up of lots of different kinds of numbers.

We have the natural numbers, or counting numbers, such as

$$1, 2, 3, \ldots$$

These are numbers for which addition and multiplication will always result in a natural number. But when we try to do some kinds of subtraction, as in the case of $3 - 9$, or $1 - 1$, we need more numbers in our set. So, we have to add zero to our collection to get the whole numbers, and we have to add negative numbers to get the integers:

$$\ldots, -3, -2, -1, 0, 1, 2, 3, \ldots$$

Now we have enough numbers for addition, multiplication, and subtraction, but not for division, so we need other numbers called ratios of integers, better known as fractions:

$$\frac{1}{2}, \frac{2}{3}, \frac{34}{37}, \text{etc.}$$

All these numbers together—the natural numbers, integers, and ratios of integers—are known as rational numbers.

Not all numbers are rational. For example, in the equation $x^2 - 2 = 0$, when we solve for x, we get $x = \sqrt{2}$. This is irrational because it doesn't work out to be something neat like $x = \sqrt{9}$, which is $x = 3$. The square root of 2, $\sqrt{2}$, is a nonrepeating, nonterminating decimal. So are numbers like π and e. We call all numbers of this type irrational.

So, the real numbers are made up of all rational and irrational numbers. No number can be both rational and irrational, so you will have at least two separate areas in your Venn diagram. A Venn diagram is a picture that is used to illustrate intersections, unions, and other operations on sets. Here is an illustration of what a Venn diagram of the different sets of numbers looks like:

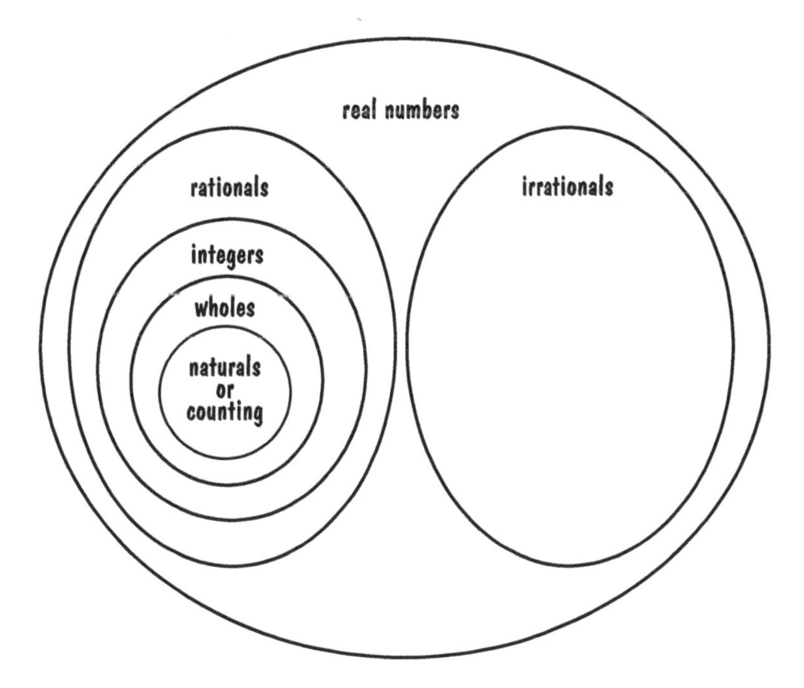

—*Dr. Math, The Math Forum*

Real Number Terminology

Dear Dr. Math,

What does it mean to be nonexistent over the reals? Can you explain this so that I can understand? I'm only in the seventh grade. Thank you.

Sincerely,

Carissa

Dear Carissa,

This idea is difficult for youngsters (or oldsters!) to grasp. The real number system, sometimes called "the reals" for short, consists of all the whole numbers (0, 3, 7, etc.), all the fractions ($\frac{1}{2}$, $-\frac{22}{7}$, etc.), and all the decimal fractions, even with infinitely many digits (0.72, 3.333333333 . . . , –24.380783 . . . , and the like, including π = 3.14159265358979 . . .). They are called *real* numbers because they are used to measure real things like lengths, speeds, volumes, areas, forces, and so on.

Often we are faced with the task of finding a real number that is a solution in an equation, such as the solution x in the equation $7x - 13 = 8$, or the solution a in $a^2 = 81$. Students are taught how to deal with these equations in algebra class.

Sometimes, however, there is no real number that works for a given equation. The usual example is $x^2 = -1$. Why is there no real number x that works in this equation? It is because if x is positive, x^2 will be positive; but if x is negative, x^2 will still be positive; and if x is zero, x^2 will be zero. This covers every possible real number, since each one is either positive, negative, or zero. So x^2 is positive or zero, but –1 is negative, so the two can never be equal within the set of real numbers. The phrase that is used to express this situation is that a solution x does not exist in the real number system (or "over the reals").

Sometimes, however, we really do want to have a solution to $x^2 = -1$. This can be very useful in higher mathematics for reasons too complicated to go into here. Since no real number will do, we have to invent a new kind of number, called an *imaginary number.* This name was chosen to contrast it with a real number.

The imaginary number that is the solution of $x^2 = -1$ is called i, and it is also called the square root of –1. The creation of this new kind of number allows us to expand our number system by looking at all numbers of the form $a + b \cdot i$, where a and b are real numbers, and i is the square root of –1. This system of numbers is called the *complex number system.*

The real numbers are part of this system because they can be expressed as **complex numbers** for which b is zero—that is, $a + 0 \cdot i = a$. In the complex number system, every number has a square root, which isn't true in the real number system (as we saw above).

—Dr. Math, The Math Forum

Resources on the Web

Learn more about real numbers at these sites:

Math Forum: Dr. Math Elementary Archives: About Numbers

mathforum.org/library/drmath/sets/elem_number_sense.html

A collection of letters with information about numbers, number sense, and number classifications.

Math Forum: Dr. Math Middle School Archives: About Numbers

mathforum.org/library/drmath/sets/mid_number_sense.html

A collection of letters with information about numbers, number sense, and number classifications.

Math Forum: Pascal's Triangle

mathforum.org/workshops/usi/pascal/pascal_intro.html

Some history, an interactive exploration of number patterns, and an exploratory Internet web unit for elementary, middle school, and high school, with lessons, number pattern studies, and links to relevant sites on the Web.

Math Forum: Problems of the Week: Middle School: Factors, Factoring, Prime Numbers

mathforum.org/library/problems/sets/middle_factors-primes.html

Students deepen their understanding of factors, multiples, prime factorization, and relatively prime numbers, using them in problem solving and to practice mathematical reasoning.

Math Forum: Srinivasa's Spheres (English) and Esferas de Srinivasa (Spanish)

mathforum.org/alejandre/mathfair/spheres.html

In this activity from Frisbie Middle School's Multicultural Math Fair, students work through exercises with circles and pi.

Math Forum: Teacher2Teacher FAQ: Pi Day

mathforum.org/t2t/faq/faq.pi.html

Answers to a frequently asked question: What activities can I use for Pi Day?

Math Forum: Understanding Algebraic Factoring

mathforum.org/alejandre/algfac.html

Students use algebra tiles to explore algebraic factoring.

Shodor Organization: Project Interactivate: Venn Diagrams

shodor.org/interactivate/activities/vdiagram

Students learn about classifying numbers into various categories through answering questions about Venn diagrams.

PART 4 Equations with Variables

What makes algebra interesting and worth learning? Why do we bother translating perfectly good descriptions of situations into equations?

Here's one way to think about it. Let's look at a system of equations:

$$3y + 2x = 29$$
$$2y + 5x = 17$$

There are two standard techniques (called *substitution* and *elimination*) that can be used to solve problems like this pretty easily. And here is the interesting part: once you can solve this kind of problem, you can also solve any other problem that can be set up to look like this.

For example, perhaps you're a businessperson trying to choose between two companies that want to rent you some equipment, or sell you some supplies. You might be able to reduce your problem to a pair of equations like the ones above. If so, and if you know how to solve these types of equations, then your problem is already solved, because the only thing standing between you and the solution is a technique that you already know how to apply.

So, here is the point of learning algebra: there are lots of problems in the world that look completely different from each other when expressed in English, but when translated into equations, look almost exactly the same. A relatively small set of tools—which you'll learn in algebra class—can be used to solve an infinitely large set of problems, including problems that we don't even know about yet.

In this part, Dr. Math explains

- Solving basic equations
- Graphing equations
- Equivalent equations

1 Solving Basic Equations

An equation will always have an equal sign in it. If you stop and think about it, the words *equation* and *equal* come from the same root. There are two sides to an equation, with the equal sign separating those two sides. Think of the equal sign as the middle of a balance. You can add, subtract, multiply, and divide the numbers on each side of the equation, but you always have to maintain the balance. In order to do that, whatever operation you do on one side, you have to do on the other side.

Basic Tips on Solving for *x*

Dear Dr. Math,

The following problems have tripped me up. I can't seem to figure out how to solve them. I've tried several times and had my algebra teacher explain them, but I still can't get the answers.

$$3x - 11 = 6$$
$$2x - 7 = 3 + x$$
$$3 + 3x = 9 + 2x$$
$$3x - 3 = 6 + x$$

—Clive

Dear Clive,

It might help you figure out how to solve equations by first looking at how you might make an equation. For example, suppose I start with something like

$$x = 5$$

That's no fun, is it? There's nothing to figure out. But I can multiply both sides of the equation by 3 without changing the truth of the equation, right?

$$3x = 15$$

And there are a lot of different ways I can write 3:

$$\begin{aligned} 3 &= 1 + 2 \\ &= 7 - 4 \\ &= -5 + 8 \end{aligned}$$

So, I can change my equation to look like this:

$$3x = 15$$
$$x(3) = 15$$
$$x(7 - 4) = 15$$
$$7x - 4x = 15$$

And since I can add or subtract anything on both sides of an equation without changing the truth of the equation, I can add 4x to both sides:

$$7x - 4x = 15$$
$$7x - 4x + 4x = 15 + 4x$$
$$7x = 15 + 4x$$

And, of course, I can do the same thing with any number I want:

$$7x = 15 + 4x$$
$$7x - 5 = 15 + 4x - 5$$
$$7x - 5 = 15 - 5 + 4x$$
$$7x - 5 = 10 + 4x$$

Now I have something that looks just like your problems, don't I?

The trick, then, is to reverse the steps that I just did to get back to the original, simple, boring equation:

$$7x - 5 = 10 + 4x$$
$$7x - 5 + 5 = 10 + 4x + 5$$ Add 5 to both sides.
$$7x = 15 + 4x$$
$$7x - 4x = 15 + 4x - 4x$$ Subtract 4x from both sides.
$$\frac{3x}{3} = \frac{15}{3}$$ Divide both sides by 3.
$$x = 5$$

It's easy to forget what's really going on when you're solving an algebraic expression for a variable. You're really just trying to rewrite a complicated expression in a simpler way without doing anything to change the meaning of the expression.

Now let's look at one of your example problems:

$$3 + 3x = 9 + 2x$$

The first thing to notice is that we have numbers on both sides. We know that someone put them there by adding a number to both sides of the equation, so we can take them away by subtracting a number from both sides:

$$\begin{aligned} 3 + 3x &= 9 + 2x \\ 3 + 3x - 3 &= 9 + 2x - 3 \\ (3 - 3) + 3x &= (9 - 3) + 2x \\ 3x &= 6 + 2x \end{aligned}$$

Now we have multiples of x on both sides. We can do the same kind of thing to fix that:

$$\begin{aligned} 3x &= 6 + 2x \\ 3x - 2x &= 6 + 2x - 2x \\ (3 - 2)x &= 6 + (2 - 2)x \\ x &= 6 \end{aligned}$$

Can you follow the same kinds of steps to solve the rest of the problems?

—Dr. Math, The Math Forum

Solving an Equation with Two Variables

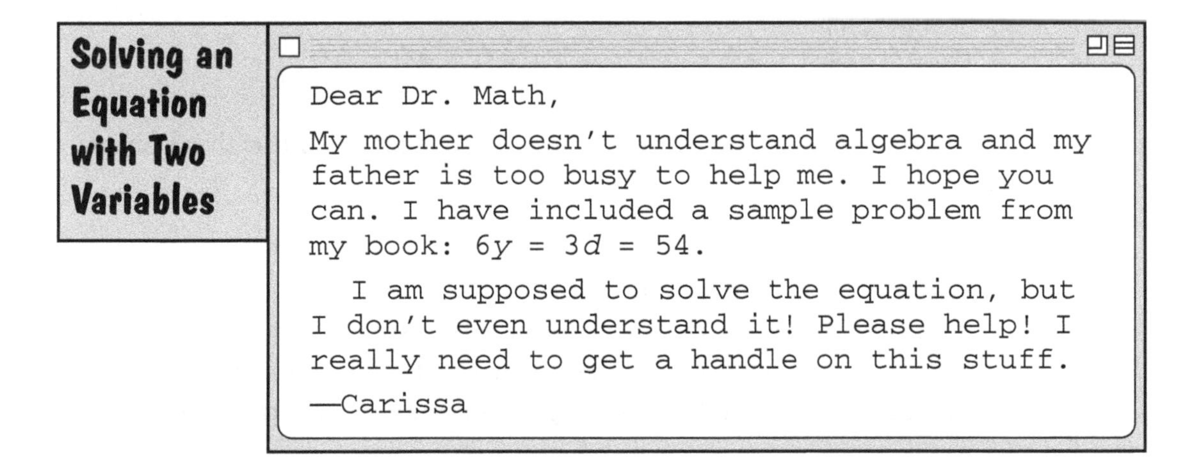

Dear Dr. Math,

My mother doesn't understand algebra and my father is too busy to help me. I hope you can. I have included a sample problem from my book: $6y = 3d = 54$.

I am supposed to solve the equation, but I don't even understand it! Please help! I really need to get a handle on this stuff.

—Carissa

Dear Carissa,

You sound really frustrated. Algebra makes a lot of people confused, especially at the beginning, but there's no magic to it. You really can learn it! I'll try to help by talking about algebra in general, and your sample problem in particular.

Algebra is like solving problems of addition, subtraction, multiplication, and division. But in algebra, instead of using a specific number, like 5, we use a letter, like *y*, to represent a number we don't know yet.

We also leave out the multiplication sign when we write equations if we're multiplying a number by a letter. What $6y$ really means is "six times *y*, which is a number I don't know yet."

The goal of almost every algebra problem is to find out what number (or numbers) the letter could equal. The math books usually call it "solving for *y*," and you've done it when your equation finally looks like "y = some number."

Your example is

$$6y = 3d = 54$$

This looks like a pretty funny equation, doesn't it? That's because . . . surprise! It's really three entirely different equations combined into a single line. They are

$$6y = 3d$$
$$3d = 54$$
$$6y = 54$$

Does this make a little more sense? Let's start with the third one:

$$6y = 54$$

In words, that's "6 times *y* equals 54."

The goal is to figure out what number we can plug in for *y* that works. You might just be able to guess the answer because you remember your times tables. To get the answer using algebra,

$$6y = 54$$

$$\frac{6y}{6} = \frac{54}{6}$$

$$y = \frac{54}{6}$$

$\frac{54}{6} = 9$, so we have $y = 9$. Great!

For $3d = 54$, you do the same thing, this time dividing both sides by 3:

$$\frac{3d}{3} = \frac{54}{3}$$

$$d = \frac{54}{3}$$

$$d = 18$$

We can use $6y = 3d$ to check our work. If you substitute 9 for y and 18 for d, is the equation true?

—Dr. Math, The Math Forum

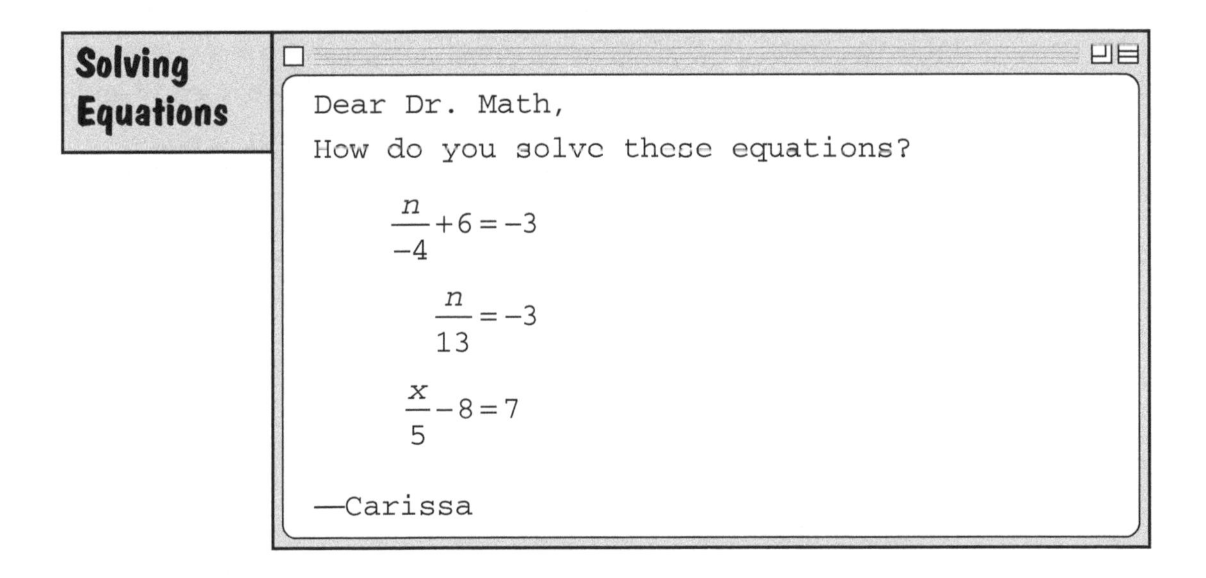

Solving Equations

Dear Dr. Math,

How do you solve these equations?

$$\frac{n}{-4} + 6 = -3$$

$$\frac{n}{13} = -3$$

$$\frac{x}{5} - 8 = 7$$

—Carissa

Dear Carissa,

Notice that in all three of these equations there is a number underneath a variable. Let's take a look at how to solve an equation of this type.

First, we have to understand what the equation is telling us. If we see something like

$$\frac{x}{3}$$

we understand that to mean "x divided by 3." If we see something like

$$\frac{1}{3}x$$

we understand that to mean "one-third times x." Why do I mention this? Because *these two are the same thing!* This is very important for you to understand. Dividing by 3 and multiplying by $\frac{1}{3}$ are the same thing.

To solve any equation, we want to end up with our variable all alone on one side, as in $x = 10$, for example. So, if we see any other numbers along with our variable, we need to perform an operation (add, subtract, multiply, or divide) to get the variable by itself.

In my example of the two ways to write x divided by 3, to get x by itself we need to do the same thing in both cases (because they are really two ways of writing the same thing). We need to multiply by 3.

Why? Because $\frac{1}{3}$ times 3 equals 1. When a variable is by itself, it has a 1 in front of it, but we never write that 1 because 1 times anything is itself. So, $1x$ is the same as x. But $(\frac{1}{3})x$ is $\frac{1}{3}$ multiplied by x. If we multiply $\frac{1}{3}$ by 3, we get 1, so if we multiply $\frac{1}{3}x$ by 3, we get $1x$ or just x.

The same is true if it is written $\frac{x}{3}$ or x divided by 3. We multiply by 3, which *cancels* the 3 on the bottom.

Do you see now why I said it was important to understand that $(\frac{1}{3})x$ and $\frac{x}{3}$ are the same?

It is also important to remember that what we do to one side of the equation, we must do to the other. If we don't, then we have put

the equation out of balance—the two sides wouldn't be equal any-more and it wouldn't really be a true equation. So, if I had

$$\frac{x}{3} = 4$$

I would multiply both sides by 3, which gives me

$$x = 12$$

If I had something else on the left side with x, like

$$\frac{x}{3} + 2 = 4$$

I would have to subtract 2 from both sides to get x by itself, like this:

$$\frac{x}{3} + 2 - 2 = 4 - 2$$

which is

$$\frac{x}{3} = 2$$

Then I would multiply by 3 to get

$$x = 6$$

I hope this helps you to solve your three equations. If you need more help, the doctor's office is open—please write back.

—Dr. Math, The Math Forum

Solving an Equation in One Variable

Dear Dr. Math,

I am reviewing for a test. Could you go step by step with me so that I don't miss anything?

$$2(x + 4) - 1 = 3 + 4(x - 1)$$

I have tried to get this question right, but the signs and the numbers keep fooling me.

—Clive

Dear Clive,

If you're having trouble keeping track of your pluses and minuses, the best way to approach these problems is to write each step (and do it really neatly) right underneath the previous one. Try not to skip steps or take shortcuts.

$2(x + 4) - 1 = 3 + 4(x - 1)$	Here's the original problem.
$(2 \cdot x) + (2 \cdot 4) - 1 = 3 + (4 \cdot x) + (4 \cdot -1)$	I use the distributive property to get the x's out in the open!
$2x + 8 - 1 = 3 + 4x - 4$	I multiply everything out.
$2x + 7 = 4x - 1$	I simplify by adding or subtracting the numbers without x terms in them.
$2x + 7 - 4x = 4x - 1 - 4x$	I subtract 4x from each side in order to move the x terms to the same side of the equation.
$2x - 4x + 7 = 4x - 4x - 1$	I apply the commutative property to move things around so that the x terms are next to each other.
$-2x + 7 = -1$	I subtract to simplify the x terms.
$-2x + 7 - 7 = -1 - 7$	I subtract 7 from each side to move the remaining numbers without x terms to the same side of the equation.
$-2x = -8$	I do the subtraction and
$2x = 8$	I multiply each side by –1 to get rid of the minus signs.
$\frac{x}{2} = \frac{8}{2}$	I divide each side by 2 to solve for x.
$x = 4$	Ta Da!

—Dr. Math, The Math Forum

Two-Step Equations

Dear Dr. Math,

I usually get these things pretty well, but I have trouble with two-step equations and the ones with variables on both sides. Can you help me solve problems like $5 - 2n = 8n - (-4n)$?

I know that problem may not work out too easily, but I just made it up off the top of my head. I have asked many people for help, but I can't get it to work out right.

Thanks,

Carissa

Dear Carissa,

In problems like the one you sent me, start by putting all the terms with a variable on one side of the equation and all the terms without a variable on the other side. If we do that to your problem, we get this:

$5 - 2n = 8n - (-4n)$	First, cancel the two negatives.
$5 - 2n = 8n + 4n$	Now add the two things on the right together.
$5 - 2n = 12n$	Now we want to get all the terms that have an n in them on one side. Add $2n$ to both sides.
$5 - 2n + 2n = 12n + 2n$	Simplify the left and right sides.
$5 = 14n$	

On one side we've just got a number, and on the other side we've got the term with the variable. The only thing left to do is get rid of the 14. To do that, we divide by 14:

$\frac{5}{14} = \frac{14n}{14}$ Cancel the 14's on the right side.

$\frac{5}{14} = n$ That's the answer!

Here's something that might help you solve equations like this. Did you notice that in the first part of solving this problem there was a $-2n$ on the left side and we turned it into a $+2n$ on the right side?

Well, that's the way it always works. If you've got something on one side of the equation and you want to bring it over to the other side, it ends up being the negative of what it was on the first side. If you had

$$y + 5 = x$$

you could turn it into

$$y = x - 5$$

The same kind of thing works for multiplication and division. If you've got

$$\frac{5}{x} = 3$$

you can turn it into

$$5 = 3x$$

When the variable was on the left side, it divided another quantity. Now that we've moved it to the right, it multiplies another quantity. But be careful to make sure that when you do this, the *whole side* needs to be multiplied or divided by the thing you're moving. For example, this doesn't work:

$$\frac{5}{x} + 4 = 7$$

$5 + 4 = 7x$ No, no, no, no, no, no, no, no!

But this does work:

$$\frac{5}{x} + 4 = 7$$

$$5 + 4x = 7x$$

And so does this:

$$\frac{4 + y}{n} = c$$

$$4 + y = cn$$

—Dr. Math, The Math Forum

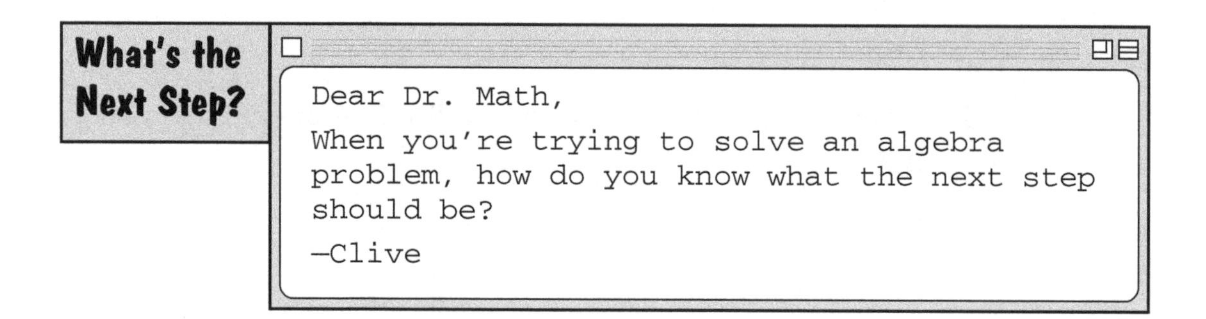

Dear Clive,

Experienced problem solvers know what steps to take next because they've got the entire solution visualized ahead of time (or at least an outline of it). They know when one equation is "simpler" than another, in the sense that it's closer to a solution.

Algebra is about learning different techniques that you can use to solve problems. Once you know the techniques, you can then use each of them to change your current problem into one that you already know how to solve . . . because you know how to change *that* problem into one you already know how to solve, . . . and so on.

At first, you learn to deal with equations in which you have a variable on one side and numbers on both sides, for example,

$$x - 4 = 7$$

Later, you learn to take an equation in which the variable has a coefficient and turn it into an equation that you already know how to solve:

$3x - 12 = 21$	Divide by the coefficient of x.
$x - 4 = 7$	No sweat! We know what to do now.

Still later, you learn to take an equation with fractions and turn it into an equation in which the variable has an integer coefficient:

$\frac{3x}{5} - \frac{24}{10} = \frac{84}{20}$	Multiply by a common denominator.
$12x - 48 = 84$	Hey, this looks familiar . . .

Then you learn that variables can appear on both sides of the equation. So, you learn to use the distributive property to combine them and end up with something you've dealt with before:

$5x - 4 = 7 + 2x$	Subtract the smaller variable term from both sides.
$5x - 2x - 4 = 7 + 2x - 2x$	Cancel.
$5x - 2x - 4 = 7$	Use the distributive property.
$(5 - 2)x - 4 = 7$	Simplify.
$3x - 4 = 7$	Now we know where we are!

And on it goes.

The key is to be able to recognize how to turn the problem you have into one that is somehow "closer" to a problem that you know how to solve. And how do you develop this ability? By solving lots of problems . . . and by getting used to the idea that most of the time what you're looking for isn't the answer but a simpler problem.

—Dr. Math, The Math Forum

2 Graphing Equations

Remember the information we covered in Part 2 on graphing points on a graph with an x- and y-axis? Using that information will help you with graphing linear equations—in other words, making graphs that are straight lines. You can graph a linear equation if you know just two coordinate pairs that make the equation true. For example, if you had the equation $2x = y$, you know that (0, 0) and (5, 10) would satisfy the equation. If you graphed those two coordinate pairs and then connected them using a straight edge and putting arrows on the ends of your lines, you would have a graph of the linear equation $2x = y$.

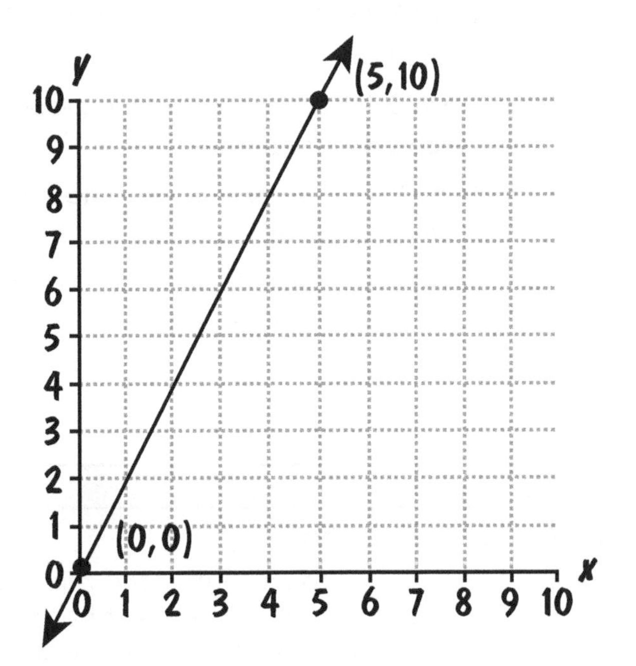

What Is a Quadrant?

Dear Dr. Math,

What is a quadrant?

—Carissa

Dear Carissa,

Quadrant means "quarter" and in particular one of the four parts into which a plane is divided by the coordinate axes. Quadrants are numbered from 1 to 4, starting at the x axis and proceeding counterclockwise (since we measure angles in that direction):

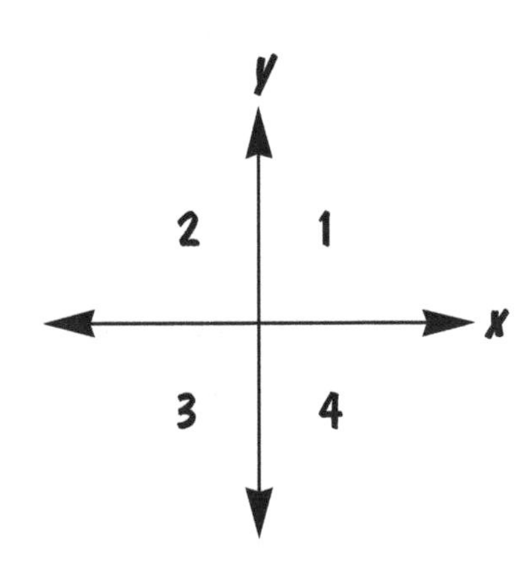

Note that you can tell which quadrant a coordinate pair would be in by looking at the signs. Here are some examples:

Quadrant 1: (5, 9) or (8, 3)—the x-coordinate and y-coordinate are positive.

Quadrant 2: (–5, 9) or (–8, 3)—the x-coordinate is negative and the y-coordinate is positive.

Quadrant 3: (–5, –9) or (–8, –3)—the x-coordinate and y-coordinate are negative.

Quadrant 4: (5, –9) or (8, –3)—the x-coordinate is positive and the y-coordinate is negative.

—Dr. Math, The Math Forum

Graphing Equations

Dear Dr. Math,

I need to find out how to graph the following equations. I know the coordinates but would like to know the name of the kind of graph I should make and where I can graph them.

They are as follows:

$$4x - 2y = 5$$
$$x = 3y - 1$$

and

$$y = 2x + 7$$
$$y = -x - 3$$

Your help would be greatly appreciated!
—Clive

Dear Clive,

All of your examples are *equations of a line,* which means that they all can be put into the form

$$ax + by + c = 0$$

The last two equations have the form $y = m \cdot x + b$. These are often called the *slope-intercept* form of a line because m is the slope of the line and b is the y-intercept, where the line crosses the y-axis.

To graph any one of these equations, all you need are two points that satisfy them, since a line is completely determined by two of its points.

Take the equation $4x - 2y = 5$, for example. To find two points on a line that satisfy this equation, first let $x = 0$ in the equation: $4 \cdot 0 - 2y = 5$. Since $4 \cdot 0 = 0$, this is the same as saying $-2y = 5$, so $y = -\frac{5}{2}$, and $(0, -\frac{5}{2})$ is a point on the line. Next, let $y = 0$ in the equation and solve for x. $4x - 2 \cdot 0 = 4x - 0$, so $4x = 5$ and $x = \frac{5}{4}$. This gives $(\frac{5}{4}, 0)$ as a second point on the line. Now just find these two points on your graph and draw a line that intersects them.

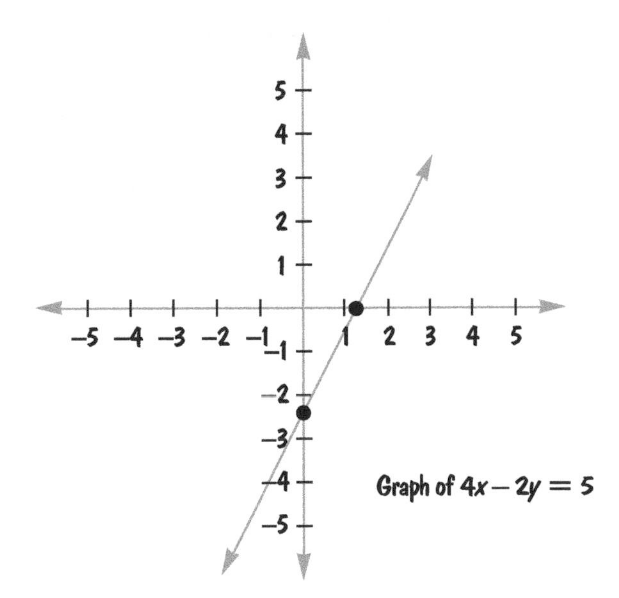

You can do the same with the equation $y = 2x + 7$: $(0, 7)$ and $(-\frac{7}{2}, 0)$ are two points on the line. Or you can notice that the slope of the line is 2 and its y-intercept is 7, because the equation is in the form $y = mx + b$. The 2 is in front of the x, which means that 2 is the slope, and the 7 is added at the end, which means that 7 is the y-intercept. If you know about slope (see box on the next page), you can plot the line with this information.

—Dr. Math, The Math Forum

Slope can be described as rise over run—how much a line rises vertically over how much it runs horizontally. (If the slope is negative, that means the line has a negative rise, which is a drop!) So, to graph using a point and slope, first make sure the slope is in fraction form. Then find the point on your graph, count up from that point by the numerator of the slope fraction, and over by the denominator. Mark your second point. You could draw your line between those two points, but plotting a third point and making sure it's on the same line is a good way to check your work.

Finding *y*-Intercepts

Dear Dr. Math,

I'm having a problem with finding *y*-intercepts. How do you know what the *y*-intercept is for $2x + y = 3$ or $x - 4y + 8 = 0$?

I'm also having trouble with this: Draw a line where the *x*-intercept = 2 and the slope = $\frac{3}{4}$.

—Carissa

Hi, Carissa,

The *y*-intercept is the place where the graph hits the *y*-axis. If you think about what the *y*-axis is, it represents all the points on the graph where the x value is zero, right?

Putting this another way, to find a point given the x- and *y*-coordinates, you move left or right by the x amount, then up or down by the *y* amount. To stay on the *y*-axis, you make no motion in the x direction; in other words, the x-coordinate is zero.

So, the *y*-intercept of $2x + y = 3$ occurs when $x = 0$, or when $2 \cdot 0 + y = 3$, or when $y = 3$. To find the *y*-intercept for $x - 4y + 8 = 0$, set $x = 0$, making $-4y + 8 = 0$, or $4y = 8$, or $y = 2$. That's all there is to it.

For your second problem, to draw a line with an x-intercept of 2 and a slope of $\frac{3}{4}$, you know one point already—on the x-axis at the point 2.

A slope of $\frac{3}{4}$ means that for every 4 units you move in the x direction, the line goes up 3 units in the y direction. So, starting from the one point you know—at (2, 0)—move to the right 4 units and up 3 units to the point (6, 3), and that's another point on the line. Just draw the straight line through those points.

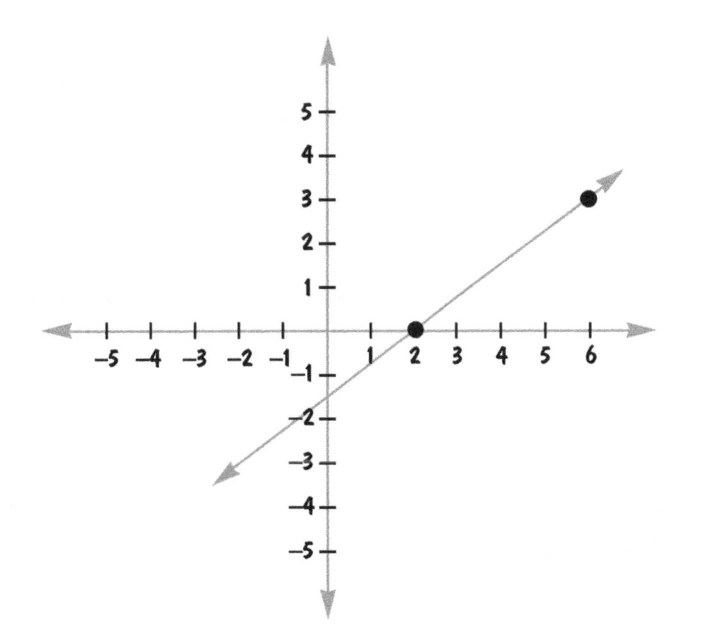

To convince yourself that the slope concept makes sense, go over another 4 units and up another 3 units to (10, 6), then do the same to (14, 9), and notice that all those points lie on the line.

I think it's very important to have this sort of a geometric or visual idea of a slope. The numerator represents the motion in the y direction, and the denominator represents the motion in the x direction. If the slope is 5, that's the same as $\frac{5}{1}$—numerator of 5 and denominator of 1. And negative slopes simply indicate that instead of sloping up, the line slopes down as you move to the right.

—Dr. Math, The Math Forum

EQUATIONS FOR GRAPHING

You can write equations in a lot of different ways. What's nice about the form

$$y = mx + b$$

is that it allows you to sketch the graph of the corresponding line very quickly.

How do you do that? You know that the line (if it's not vertical) has to cross the y-axis somewhere . . . in particular, where $x = 0$. Since you know $x = 0$,

$$\begin{aligned} y &= m(0) + b \\ &= b \end{aligned}$$

which tells you that $(0, b)$ is a point on the line. You also know that the line (if it's not horizontal) has to cross the x-axis . . . in particular, where $y = 0$. Since you know $y = 0$,

$$\begin{aligned} 0 &= mx + b \\ -b &= mx \\ -\tfrac{b}{m} &= 0 \end{aligned}$$

which tells you that $(-\frac{b}{m}, 0)$ is another point on the line. So, just by looking at the equation

$$y = mx + b$$

you can graph two points:

$$(0, b) \text{ and } (-\tfrac{b}{m}, 0)$$

And once you've graphed these two points, you can fill in the rest of the graph by drawing a line through these two points with a ruler.

Another nice thing about the form

$$y = mx + b$$

is that you can tell pretty quickly whether two lines are parallel or perpendicular. If they have the same slope, they are parallel. If the slopes can be multiplied to get −1, they are perpendicular. For example, the following two lines are parallel:

$$\begin{aligned} y &= 2x + 3 \\ y &= 2x + 14 \end{aligned}$$

And the following two lines are perpendicular:

$$\begin{aligned} y &= 2x + 3 \\ y &= -\tfrac{1}{2}x - 6 \end{aligned}$$

Equivalent Equations

Two equations are **equivalent** if the truth of either implies the truth of the other. For example, the equations

$$3y + 2x = 5$$

and

$$6y + 4x = 10$$

are equivalent, even though the expressions $3y + 2x$ and $6y + 4x$ are not equivalent expressions.

Why is this important? Sometimes you want to transform an equation into an equivalent equation because that will make a problem simpler to solve. It's the same sort of idea as transforming a fraction like $\frac{3}{8}$ into an equivalent fraction like $\frac{9}{24}$ so that you can add it to a third fraction like $\frac{7}{24}$.

The key idea is to distinguish between *meaning* and *appearance*. Many of the techniques that you'll learn in algebra will focus on changing the appearance of an equation without changing the meaning (or truth) of the equation. It's a little bit like getting change for $20—if you can get a ten, a five, three ones, and eight quarters, you can lend $3 to your friend for a movie and use the quarters later to play a video game. It's the same amount of money, but changing it to a different form can make it more useful.

Rx EQUIVALENT EXPRESSIONS

Equivalent expressions have the same value for every possible value of the variables they include. $5 \cdot (x - 4)$ and $5 \cdot x - 20$ are equivalent expressions because no matter what the value of x, these expressions are equal in value.

Subtracting One Equation from Another

Dear Dr. Math,

If you have two equations with two variables—the same variables in each equation, say x and y—there are two generally accepted ways of solving the equations. You can isolate one variable in one equation, then substitute it in the second equation. Or you can manipulate one of the equations so that you have equal values of one variable, then subtract one equation from the other, eliminating the variable. I can do the subtraction and solve the equations, but no one ever answered this question:

Why can you subtract one equation from another? Not why can you subtract one mixed variable from another, for example, $5x - 3x$ or $9x^2 - 5x^2$. That's clear enough. But why can you subtract one equation from another and get something meaningful that you can use to solve the equations? The explanations I've heard are just given in more math jargon, which is no help.

I have another question. If you graph the equations, you get two lines. How can you subtract a line from another line? You can subtract the length of one line from the length of another, but how can you subtract lines? Lines are not numbers; they're marks on pieces of paper or chalkboards. There are points all along the lines, but no one says, "Subtract every point on the second line from every point on the first line." I still don't get it.

—Clive

Hi, Clive,

Great questions! Let's try an example. We want to solve

$$2x + 5y = 26$$
$$x - 3y = 2$$

We double the second equation, giving

$$2x - 6y = 4$$

then subtract the equations to get

$$0 + 11y = 22$$

so $y = 2$.

Why can we subtract one equation from another? If you picture an equation as a balanced scale, this is easy to explain: take the same amount off both sides of the scale, and it remains balanced. Here, $2x - 6y$ is the same as 4, so we can subtract 4 from each side of the first equation (which will not change its meaning) by subtracting $2x - 6y$ from the left side and 4 from the right. So, the new equation will be true whenever both original equations are true. This gives us a simpler equation we can solve in order to find the solution to the original pair.

Technically, we want to transform the problem into an equivalent one—that is, one with exactly the same set of solutions—but the new equation by itself is not equivalent to both original equations. Although any solution to the original pair will make this equation true, not all solutions to the new equation will be solutions of the pair. You can see this easily here, where x doesn't even appear in the new equation, so it can be anything. What's really happening is that the new equation *together with* one of the pair is equivalent to the original pair.

Now, what does this mean in terms of the graphs of the equations? Not much. Notice that if we multiply an equation by 2, as we might do before subtracting, we haven't changed its meaning at all—it's still the same line. So, this action on the equation has no effect on the

graph. Since we can subtract any such form of the equation of one line from the other, you can see that the line itself doesn't determine what will happen when we subtract equations. I can't draw two lines and then say, "This third line is the difference between those two lines," because there are many different possibilities for the result, depending on which equation we used for each line. In other words, you can't talk about subtracting two lines in this sense; you can only subtract equations.

—Dr. Math, The Math Forum

Resources on the Web

Learn more about equations with variables at these Math Forum sites:

Algebraic Thinking

mathforum.org/alejandre/frisbie/thinking.html

Students use algebraic thinking to determine which variables stand for which digits in a series of simple equations.

Lewis Howard Latimer

mathforum.org/alejandre/mathfair/latimer.html

Students work through a mathematical investigation using the work of Lewis Howard Latimer.

Linear Equation Activity

mathforum.org/workshops/sum98/participants/ader/LinEq_les1.html

Learn to set up a problem and develop a formula by working through two examples.

Problems of the Week: Algebra: Linear Equations

mathforum.org/library/problems/sets/alg_linear.html

Algebra problems that involve linear equations.

Setting Up a Problem on a Spreadsheet

mathforum.org/workshops/sum98/participants/sinclair/problem/setting.html

Given a real world situation, students are asked to write an equation that represents the problem.

Algebra Applications

After reading the preceding parts, you might still be wondering about the uses of all these numbers, equations, and variables.

We gave you a hint in Part 4, when we noted that much of the power of algebra comes from its ability to express all kinds of different problems in a way that makes them look very similar to each other—similar enough that the same set of tools can be applied to many types of problems.

There's another important aspect of numbers and equations that makes them tremendously valuable: unlike objects in the real world, they're free! To go out and try to solve problems using things in the real world can be difficult and expensive. If we can get the same answers by *simulating* a world using numbers and then use that simulated world as a place to try out our ideas, we can save time, money, and effort.

Maybe the problem you're solving involves choosing the best shape for a garden, or figuring out how many pens and pencils can be purchased for a certain amount of money. You *could* solve the problems by building (and rebuilding, and rebuilding) the garden, or by getting some pens and pencils and trying out all the different combinations . . . but it's much easier to use numbers to simulate the objects and experiment with the numbers instead.

There are some very useful ways of using numbers that let you solve lots of similar problems. This part will show you four of these techniques.

In this part, Dr. Math explains

- Ratio and proportion
- Area and perimeter
- Distance, rate, and time problems
- Rate of work problems

1 Ratio and Proportion

Say you're a junior in high school, getting a jump on college selection. One thing to look for in a college is the student–teacher ratio. One school reports its student–teacher ratio as 20:1. If you attended that school, would you expect to see any classes with 40 students in them? Sure, but you'd expect them to be balanced out with either an extra teacher, or another class somewhere with fewer than 20 students, since on average there should be 20 students per teacher.

Ratios compare numbers. What if we want to compare ratios? Then we use **proportions.** True proportions tell us whether two ratios are equal. What if another school reports its student–teacher ratio, rather oddly, as 52:3? (Maybe it's a very small school with only three teachers, and it didn't want to reduce the ratio.) Is that better or worse than 20:1? You could use a proportion to find out.

Ratio and Proportion

```
Dear Dr. Math,
I need extra help with ratios and propor-
tions. I would appreciate it if you would
write back with examples.
Thanks,
Clive
```

Dear Clive,

I'm not sure what kind of examples of ratios you were thinking of, but I'll throw out a few for you to think about. Putting two numbers in a ratio gives you a sense of the relative sizes of two numbers. For instance, if I want to know the ratio of 200 to 100, I say it is $\frac{200}{100} = \frac{2}{1}$. That means 200 is twice the size of 100.

Proportionality is closely related to ratios. If I have a triangle with sides 3, 4, and 5, and I have another triangle with sides in proportion to the first triangle, the second triangle's sides must be in the same ratio to one another as those of the first triangle. So, for instance, a triangle with sides 6, 8, and 10 is proportional to the triangle of sides 3, 4, and 5, because its sides are exactly twice as big as those of the first triangle.

—Dr. Math, The Math Forum

What Is a Ratio?

Dear Dr. Math,

I need to know what a ratio is and how to do it. I have searched my whole math book. Do you think you can help me? If so, thank you very much!

Thanks,

Carissa

Hi, Carissa!

While researching your question, I realized how confusing ratios can be for someone without too much experience in mathematics. Let's see if we can clear up the confusion!

Ratios are pairs of numbers, and they are used to make comparisons. We need to be careful with the meaning behind the notation we use. A fraction is different from a ratio, although sometimes the two get confused because they look similar. The second number in a fraction is called the denominator, and it represents the number of parts in the whole. For instance, if a recipe is $\frac{3}{5}$ (a fraction) flour, for every 5 parts of the total recipe, 3 parts are flour and the other 2 are everything else. In this case, the ratio of flour to other things is 3:2.

Ratios can be written three different ways:

1. 2 to 3
2. 2:3
3. $\frac{2}{3}$

Pretend I have a class of 25 students: 10 students are boys and 15 are girls. I can compare the number of boys to the number of girls using the ratio $\frac{10}{15}$ (or 10:15, or 10 to 15). This ratio means the same thing as saying that for every 10 boys in my class, I have 15 girls. You can reduce this fraction (remember, ratios are comparisons using fractions) to $\frac{2}{3}$ by dividing both numerator and denominator by 5. So, you can also say that there are 3 girls for every 2 boys in the class.

Let's say some new kids move into the neighborhood and join my class, but we want to keep the ratio of boys to girls the same. To do this every time 2 boys join the class, 3 girls have to join as well. If more or fewer girls or boys than this join my class, the ratio of boys to girls in my class would change.

This idea of keeping the ratio the same can be important. Let's say I'm making brass in a factory. Brass is a mixture of copper and zinc. Different kinds and colors of brass result from changing the mix of the two metals. One kind of brass has 7 parts copper for every 3 parts zinc. This ratio is written as $\frac{7}{3}$. Would a mix of 14 parts copper and 6 parts zinc make the same brass? This ratio would be written as $\frac{14}{6}$. Using your knowledge of equivalent fractions, you can see that $\frac{7}{3} = \frac{14}{6}$ (multiply the numerator and denominator of $\frac{7}{3}$ by 2 and you get $\frac{14}{6}$). So, the brass is the same. If one day I didn't have enough zinc to make the ratio of copper and zinc the same as $\frac{7}{3}$, I would be making a different kind of brass, and my customers might be very upset because the color would be different.

—Dr. Math, The Math Forum

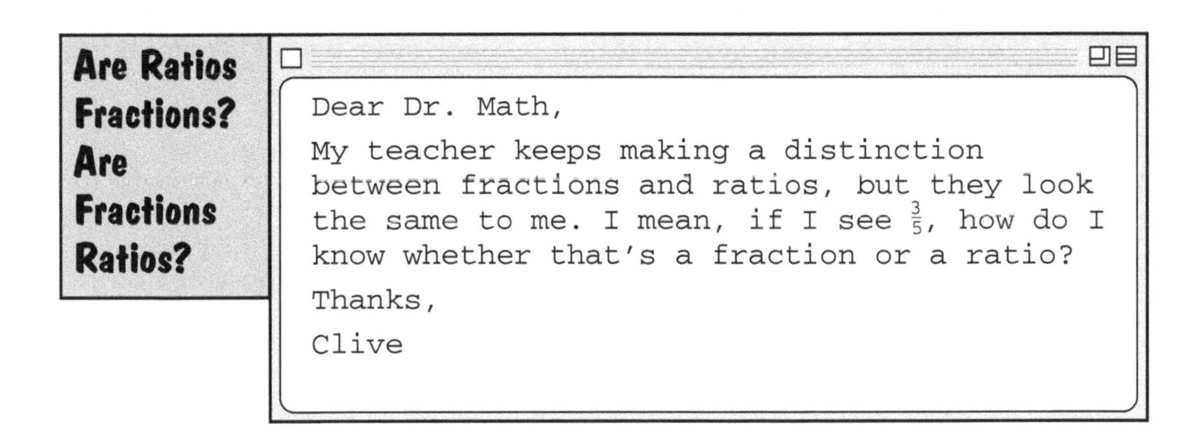

Are Ratios Fractions? Are Fractions Ratios?

Dear Dr. Math,

My teacher keeps making a distinction between fractions and ratios, but they look the same to me. I mean, if I see $\frac{3}{5}$, how do I know whether that's a fraction or a ratio?

Thanks,

Clive

Hello, Clive,

In the end, this is a problem only if you forget that fractions and ratios are both just divisions that you haven't worked out yet. So long as you set up a problem with corresponding types of quantities in corresponding places, it doesn't really matter what you call them.

For example, at a certain school, there are 300 boys and 270 girls. At another school with the same ratio of boys to girls, there are 400 girls. How many boys are at that school?

You might set this up as

$$\frac{300 \text{ boys}}{270 \text{ girls}} = \frac{? \text{ boys}}{400 \text{ girls}}$$

Or, you might set this up as

$$\frac{300 \text{ boys}}{(300 + 270) \text{ students}} = \frac{? \text{ boys}}{(? + 400) \text{ students}}$$

Some people (including your teacher) might say that one of these equations uses ratios, while the other uses fractions, but in the end they're *all* ratios. In one case, you have the ratio of the size of one subset to the size of a second subset; in the other case, you have the ratio of the size of one subset to the size of the entire set. It really doesn't matter at all what you call the ratios, so long as you have the same sorts of things on each side. If you do something like

$$\frac{300 \text{ boys}}{(300 + 270) \text{ students}} = \frac{? \text{ boys}}{400 \text{ girls}}$$

then you're in trouble, because the ratios on opposite sides of the equation no longer correspond to each other. But if you put corresponding quantities in corresponding locations, then a lot of us don't really care *what* you call it.

Both fractions and ratios express a relationship between two numbers. The important thing is to be clear on the units and the relationships between them.

—Dr. Math, The Math Forum

Ratios as Fractions

Dear Dr. Math,

I am in grade 7 and I always have trouble with my math. I am particularly stuck in the area of ratios. For example, last month we started trying to put ratios into simplest form, like 64:12 = ? Please help, Dr. Math!

Thank you very much,

Carissa

Hello, Carissa,

One way to look at 64:12 is as the fraction $\frac{64}{12}$. This is called an *improper fraction* because its numerator is bigger than its denominator, but it is still a perfectly good fraction. You have probably had problems where you were supposed to reduce a fraction to its lowest terms. An example is that the fraction $\frac{7}{14}$ equals $\frac{1}{2}$. The ratio problem is really just the same old fraction problem in different words.

If we reduce $\frac{64}{12}$, we get $\frac{32}{6}$, and we can reduce that to $\frac{16}{3}$. We can't go any further than that, so in terms of ratios, 64:12 equals $\frac{16}{3}$. I hope this helps on the mechanics of what you are supposed to do for this kind of problem.

—Dr. Math, The Math Forum

Cross-products, Ratios, and Proportions

Dear Dr. Math,

I am having trouble understanding cross-products. Will you please explain them to me?

—Carissa

Hi, Carissa!

Cross-products can be used for three similar purposes: to compare fractions, to determine whether a proportion is true, and to solve a proportion.

Fractions that represent the same quantity are called equivalent fractions. For example, $\frac{3}{6}$, $\frac{4}{8}$, and $\frac{5}{10}$ are all equivalent fractions for $\frac{1}{2}$. You can use cross-products as a shortcut to find whether two fractions are equivalent. If the cross-products are equal, then the fractions are equivalent; if the cross-products are not equal, then the fractions are not equivalent. To find the cross-product of two fractions, multiply the numerator of the first fraction by the denominator of the second. Then multiply the denominator of the first fraction by the numerator of the second.

Why do cross-products work? Well, if you have two ratios, $\frac{a}{b}$ and $\frac{c}{d}$, what do you get if you divide them?

$$\frac{a/b}{c/d}$$

Invert and multiply:

$$\frac{a/b}{c/d} = \frac{a}{b} \cdot \frac{d}{c} = \frac{ad}{bc}$$

There are three possibilities here: (1) the two ratios are equivalent (the cross-products are equal); (2) the top ratio is larger (the product ad is larger than the product bc); or (3) the top ratio is smaller (the product ad is smaller than the product bc).

For example, to find out if $\frac{3}{10}$ and $\frac{15}{50}$ are equivalent:

$3 \cdot 50 = 150$
$10 \cdot 15 = 150$, so the fractions are equivalent.

What about $\frac{7}{14}$ and $\frac{5}{8}$? Are they equivalent?

$7 \cdot 8 = 56$
$14 \cdot 5 = 70$, so the fractions are *not* equivalent.

A proportion is a statement that two ratios, usually expressed in fractions, are equivalent. You can tell a proportion is true by using cross-products to determine if the fractions are equivalent.

Is the proportion $\frac{3}{12} = \frac{8}{32}$ true or false? (We read $\frac{3}{12} = \frac{8}{32}$ like this: "3 is to 12 as 8 is to 32.") It's true if the fractions are equivalent. Let's reduce them:

$$\frac{3/3}{12/3} = \frac{1}{4}$$

$$\frac{8/8}{32/8} = \frac{1}{4}$$

So, the fractions are equivalent. We can show this in a simpler way using cross-products. Does $3 \cdot 32 = 8 \cdot 12$?

$$3 \cdot 32 = 96$$
$$8 \cdot 12 = 96$$

Yes, the proportion is true.

Cross-products can also be used to solve proportions when one of the numbers is unknown:

$$\frac{5}{20} = \frac{n}{48}$$

What is n? We know from what we just learned that $5 \cdot 48$ must equal $20 \cdot n$:

$$20 \cdot n = 5 \cdot 48$$
$$20 \cdot n = 240$$

To solve for n, divide both sides of the equation by 20:

$$240 \div 20 = 12$$
$$n = 12$$

We check this value for n by substituting it back into the proportion:

$$\frac{5}{20} = \frac{12}{48}$$

Are the two fractions equivalent?

$$\frac{5/5}{20/5} = \frac{1}{4}$$

$$\frac{12/12}{48/12} = \frac{1}{4}$$

So, the proportion is true; our value of 12 for n is correct.

—Dr. Math, The Math Forum

Area and Perimeter

There are many practical reasons for understanding area and perimeter. Suppose you're buying a new carpet. You must figure out the area of the floor that you need covered. If you decide to add a border around the carpet, then you need to know the carpet's perimeter (the measure around the carpet's edge).

Understanding Area and Perimeter

Dear Dr. Math,

I do not understand area and perimeter.

—Clive

Dear Clive,

The word *perimeter* literally means "distance around." Think about a rectangle like this one:

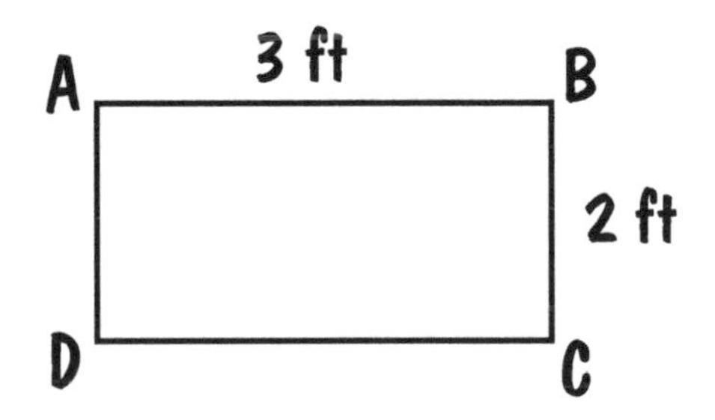

One way to walk around the rectangle would be to move from A to B (a distance of 3 ft), then from B to C (a distance of 2 ft), then from C to D (a distance of 3 ft), and finally from D to A (a distance of 2 ft). The total distance involved would be 3 ft + 2 ft + 3 ft + 2 ft, or 10 ft. So, that's the perimeter of the rectangle: 10 feet.

Area is more complicated because it involves two dimensions, whereas perimeter involves only one. The way I always think of area is in terms of the amount of paint that I would need to cover a shape.

If something has twice as much area as a smaller thing, then I'd need twice as much paint to cover the larger shape.

For a rectangle, we compute area by multiplying the length by the width:

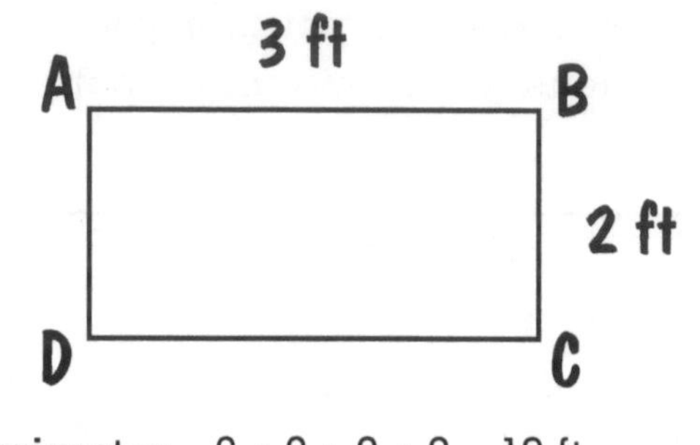

perimeter = 3 + 2 + 3 + 2 = 10 ft
area = 3 · 2 = 6 square feet

If we double the length of each side, we get twice the perimeter but *more* than twice the area:

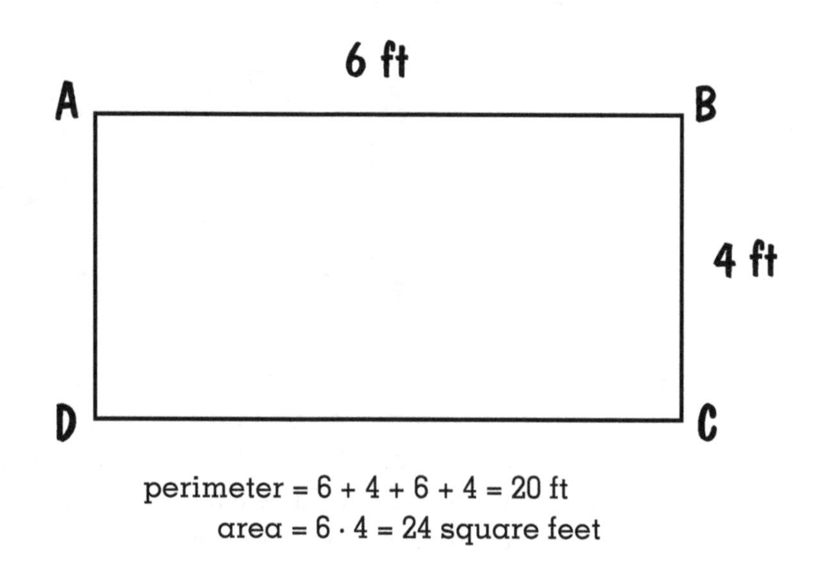

perimeter = 6 + 4 + 6 + 4 = 20 ft
area = 6 · 4 = 24 square feet

A lot of people get confused about this point, but a diagram can help make things clearer:

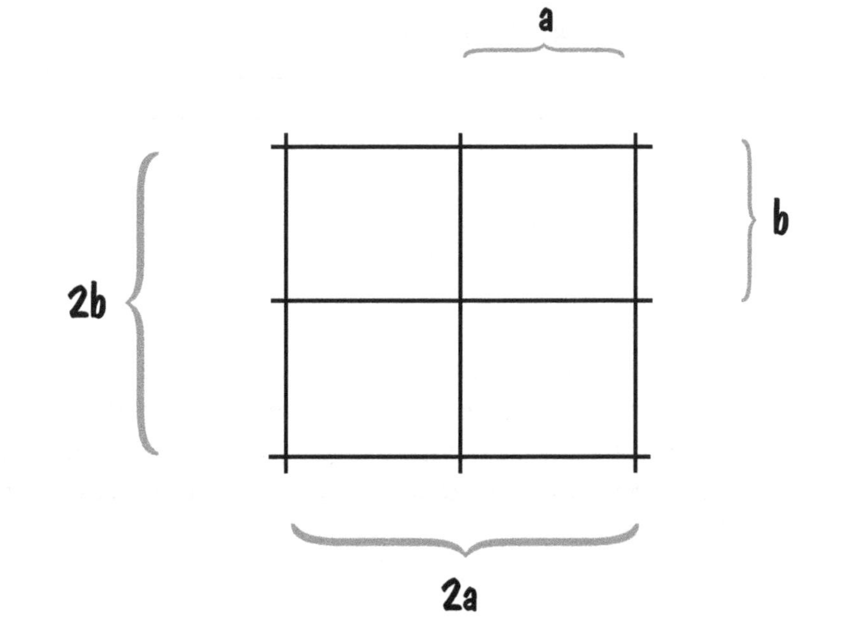

For a rectangle, if I double the length of each side, I get four times the area (but twice the perimeter). Note that we can have rectangles with the same perimeters but different areas:

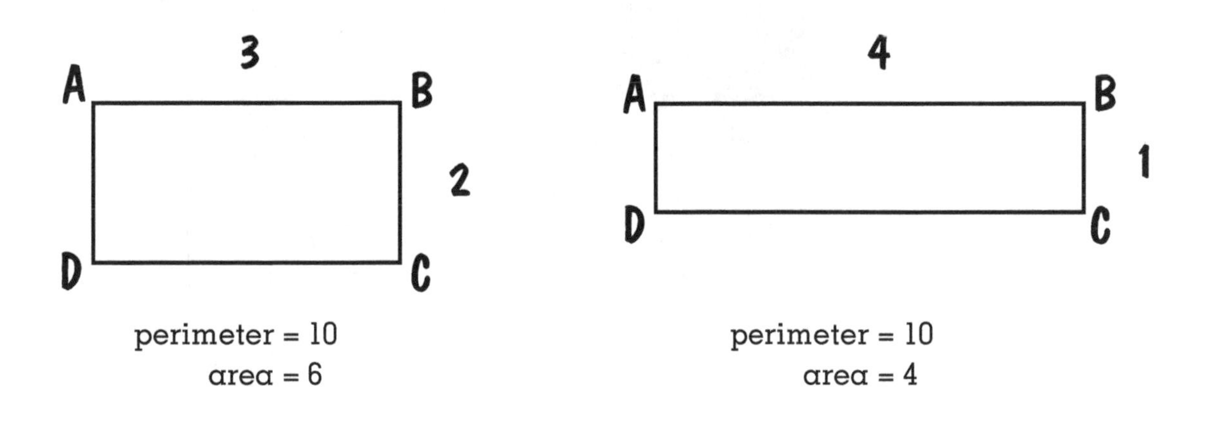

In fact, for a given perimeter, we can make the area as close to zero as we'd like by making the rectangle long and thin:

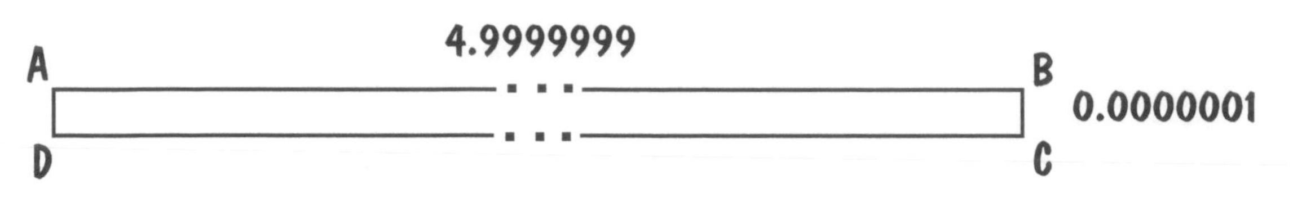

perimeter = 10
area = .0000005

For polygons (triangles, pentagons, hexagons, and other shapes that you make by linking line segments together), perimeter is always pretty easy to compute (you just add up the lengths of the sides), but computing area gets more complicated. For a triangle, the formula is

$$\text{area} = \frac{1}{2} \cdot \text{length of base} \cdot \text{height}$$

For a trapezoid, the formula is

$$\text{area} = \frac{1}{2} \cdot \text{top length} + \text{bottom length} \cdot \text{height}$$

Note that finding the height for rectangles is easy—you just choose any side to be the height and either adjacent side as the length. For other shapes, it can be a little more involved and often requires the use of the Pythagorean theorem (which is one of the reasons that your teachers want you to learn it).

—*Dr. Math, The Math Forum*

Distance, Rate, and Time Problems

The formula distance = rate · time expresses one of the most frequently used relations in algebra. When using this equation, it's important to keep the units straight. For instance, if the rate the problem gives is in miles per hour (mph), then the time needs to be in hours, and the distance in miles. Always make your units match.

ASK DR. MATH FAQ: DISTANCE, RATE, AND TIME

The formula distance = rate · time expresses one of the most frequently used relations in algebra. Since an equation remains true as long as you divide through by the same nonzero element on each side, this formula can be written in different ways:

To find **rate,** divide through on both sides by *time:*

$$\text{rate} = \frac{\text{distance}}{\text{time}}$$

Rate is distance (given in units such as miles, feet, kilometers, meters, etc.) divided by time (hours, minutes, seconds, etc.). Rate can always be written as a fraction that has distance units in the numerator and time units in the denominator, for example, 25 miles/1 hour.

To find **time,** divide through on both sides by *rate:*

$$\text{time} = \frac{\text{distance}}{\text{rate}}$$

When using this equation, it's important to keep the units straight. For instance, if the rate in the problem is in miles per hour (mph), then the time needs to be in hours and the distance in miles. If the time is given in minutes, you will need to divide by 60 to convert it to hours before you can use the equation to find the distance in miles. Always make your units match: if the time is given in fortnights and the distance in furlongs, then the rate should be given in furlongs per fortnight.

Say a car is traveling at 30 mph and you want to figure out how far it will go in 2 hours. You can use the formula

$$\text{rate} \cdot \text{time} = \text{distance}$$

$$30\,\frac{\text{miles}}{\text{hour}} \cdot 2\text{ hours} = 60\text{ miles}$$

The hours cancel, leaving only miles.

What if you want to calculate the number of miles a car traveling at 30 mph goes in 120 minutes? Since 120 minutes is equal to 2 hours (60 minutes in 1 hour · 2 hours = 120 minutes), we should get the same distance of 60 miles, but we will *not* get the answer this way:

$$30\,\frac{\text{miles}}{\text{hour}} \cdot 120\text{ minutes} = 3600\,\frac{\text{mile minutes}}{\text{hour}}$$

Now, "3600 mile minutes per hour" isn't very helpful, since we'd like our answer in miles, so we need to divide by 60 minutes per hour:

$$3600\ \frac{\text{mile minutes}}{\text{hour}} \cdot \frac{1\ \text{hour}}{60\ \text{minutes}} = 60\ \text{miles}$$

The hours and the minutes cancel, leaving only miles. However, you should always convert the minutes to hours before using the formula.

Remembering to be careful about units, let's look at a problem.

> Superheroes Liza and Tamar leave the same camp and run in opposite directions. Liza runs 2 miles per second (mps) and Tamar runs 3 mps. How far apart are they in miles after 1 hour?

To begin, we can either convert rates to miles per hour, or we can convert the time to seconds. Let's convert from miles per second to miles per hour.

There are 3,600 seconds in an hour, so if Liza runs 2 miles in a second, then she will run at $3600 \cdot 2 = 7200$ mph. Similarly, Tamar will run at $3600 \cdot 3 = 10{,}800$ mph:

$$2\ \frac{\text{miles}}{\text{second}} \cdot \frac{3600\ \text{seconds}}{1\ \text{hour}} = 7200\ \frac{\text{miles}}{\text{hour}}$$

The seconds cancel, leaving miles per hour.

Back to the problem. How far does Liza run in 1 hour? We know her rate (7,200 mph) and the time that she runs (1 hour), so we can use the formula

$$7200\ \frac{\text{miles}}{\text{hour}} \cdot 1\ \text{hour} = 7200\ \text{miles}$$

This makes sense because, by definition, if Liza's speed is 7,200 miles per hour, then she runs 7,200 miles in an hour. Tamar, whose speed is 10,800 miles per hour, will run 10,800 miles in an hour.

How far apart will the two runners be after an hour? Because they are running in opposite directions, the answer is simply the sum of the distance each runs in an hour:

$$7200 + 10{,}800 = 18{,}000 \text{ miles apart}$$

Since the earth has a circumference of about 24,000 miles at its equator, that's about three-fourths of the way around the world!

Running Rates, Line Segments, and Basic Algebra

Dear Dr. Math,

Steven ran a 12-mile race at an average speed of 8 miles per hour. If Adam ran the same race at an average speed of 6 miles per hour, how many minutes longer did Adam take to complete the race than Steven?

—Carissa

Hi, Carissa,

First, let's figure out how long it took Steven to run the 12 miles. Do you know the formula distance = rate · time? It means that the **distance** you travel is equal to the speed you are moving multiplied by how long you move at that speed.

We know that Steven ran 12 miles (distance) and that he ran at 8 mph (rate). All we have to do is to solve for the time, which turns out to be 12 (miles)/8(mph). $\frac{12}{8} = 1.5$, or 1 hour, 30 minutes.

Now we need to find out how long it took Adam to run the same 12 miles at 6 mph so that we can compare the two times and find the difference. Why don't I let you try and figure out how long it took Adam using the same method as above?

—*Dr. Math, The Math Forum*

Speed, Distance, and Time

Dear Dr. Math,

If you travel 1 mile in 45 seconds, how fast are you driving?

—Clive

Dear Clive,

Too fast! Let's figure it out.

When we are done, we want miles per hour. Right now, we have miles per second. So, if we go 1 mile in 45 seconds, the fraction is

$$\frac{1 \text{ mile}}{45 \text{ seconds}}$$

Let's change the time (denominator) to hours. First, we need to get it into minutes.

Since there are 60 seconds in a minute, we can just multiply the fraction by 60 seconds per minute. (We haven't changed the value of the fraction because 60 seconds is the same amount of time as 1 minute.)

Now we have

$$\frac{1 \text{ mile}}{45 \text{ seconds}} \cdot \frac{60 \text{ seconds}}{1 \text{ minute}}$$

One more step. To go from minutes to hours, we do almost the same thing. We multiply by 60 minutes per hour:

$$\frac{1 \text{ mile}}{45 \text{ seconds}} \cdot \frac{60 \text{ seconds}}{1 \text{ minute}} \cdot \frac{60 \text{ minutes}}{1 \text{ hour}}$$

Now we multiply all the numbers together and cancel whatever units we can, and whatever doesn't cancel are the units that we keep. In our example, the units that do not cancel are mile and hour, so those are the units that we want to keep.

The answer I get when I plug the numbers into a calculator is 80 miles per hour. As I said before, way too fast.

—Dr. Math, The Math Forum

Distance/ Time Problem

Dear Dr. Math,

Use the formula $d = rt$ to answer the following problem: Bernice is cycling around a track at 15 mph. Betty starts at the same time but only goes 12 mph. How many minutes after they start will Bernice pass Betty if the track is $\frac{1}{2}$ mile long? They are moving in the same direction.

—Clive

Dear Clive,

What matters here isn't how fast Bernice and Betty are moving relative to the *track* but how fast they are moving relative to *each other*.

Suppose they were moving in a straight line instead of around a track. After 1 hour, Bernice would be ahead by 3 miles. (She would have moved 15 miles, while Betty would have moved 12 miles.) After 2 hours, she would be ahead by 6 miles. And so on. The distance between them increases by 3 miles per hour, which is the *difference* in their speeds. It doesn't matter whether those speeds are 15 mph and 12 mph, or 115 mph and 112 mph, or 3 mph and 0 mph.

On a $\frac{1}{2}$-mile track, Bernice has to get ahead by $\frac{1}{2}$ mile to "catch up" to Betty. If she moves 3 mph relative to Betty, how long does this take? Now we're in a position to use the formula

$$\text{distance} = \text{rate} \cdot \text{time}$$

$$\frac{1}{2}\text{mile} = \frac{3 \text{ miles}}{\text{hour}} \cdot ?\text{ hours}$$

The units match up, so we can ignore them to get the equation

$$\frac{1}{2} = 3 \cdot ?$$

which we can solve by dividing each side of the equation by 3:

$$\frac{1/2}{3} = ?$$

$$\frac{1}{6} = ?$$

So, it takes $\frac{1}{6}$ of an hour, or 10 minutes, for Bernice to catch up to Betty.

—Dr. Math, The Math Forum

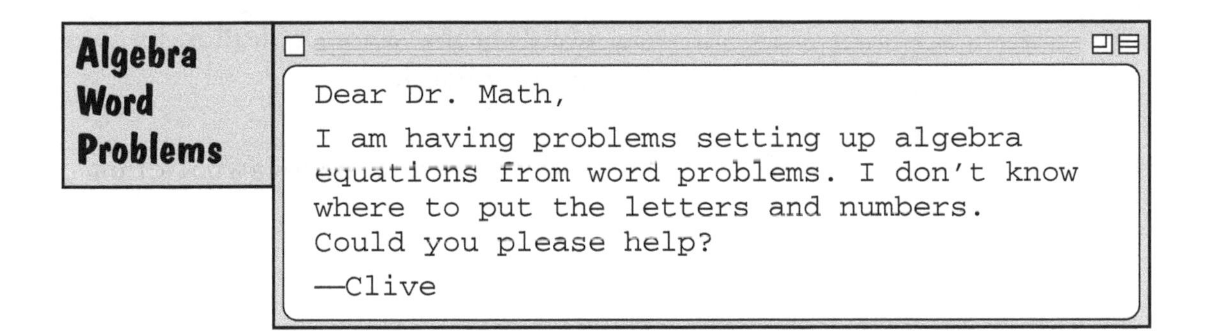

Algebra Word Problems

Dear Dr. Math,

I am having problems setting up algebra equations from word problems. I don't know where to put the letters and numbers. Could you please help?

—Clive

Dear Clive,

With word problems, you have to express the information in symbolic form, which means you use a letter to represent some quantity that you want to find or need to use in writing down an equation. For example, you could use t to represent time and x to represent distance. I will give you an example to illustrate the method, but of course, every problem will be slightly different, and there is no single way of doing word problems.

If a man walks to the station at 3 mph, he misses his train by

1 minute. If he runs at 6 mph, he has 2 minutes to spare. Find the distance to the station.

Let x = distance to the station. Now we know that the difference in *time* is 3 minutes between the two methods of travel, so we need to write down an equation involving *time*.

Time = distance/rate, so when the man walks, his total time to the station is $\frac{x}{3}$ (hours), and when he runs, his time is $\frac{x}{6}$ (hours).

We know the distance to the station is the same whether the man walks or runs. We know his two rates, and we know the difference in *time* between them. How do we write that? Rate 1 – rate 2 = difference in *time* (in hours) between them:

$$\frac{x}{3} - \frac{x}{6} = \frac{3}{60}$$

Convert x/3 into 2x/6 and subtract.

$$\frac{x(2-1)}{6} = \frac{3}{60}$$

$$\frac{x}{6} = \frac{3}{60}$$

Multiply the whole equation by 6 to get x by itself.

$$x = \frac{18}{60}$$

Reduce the fraction to lowest terms.

$$x = \frac{3}{10} \text{ of a mile}$$

So, the distance to the station is $\frac{3}{10}$ mile. You can check the result by calculating the actual time to walk and the time to run:

$$\text{Time to walk} = \frac{3/10}{3} = \frac{1}{10} \text{ hour} = 6 \text{ minutes}$$

$$\text{Time to run} = \frac{3/10}{6} = \frac{1}{20} \text{ hour} = 3 \text{ minutes}$$

Difference = 3 minutes.

The important first step is to use letters to represent the things you don't know so that you can write down equations to represent what you know. Lay out the work clearly. Keep thinking about what the problem is asking you and what you still need to find out.

—*Dr. Math, The Math Forum*

4 Rate of Work Problems

When two or more people work together on a job, they don't always work at the same speed. In order to be able to predict how long a certain job will take, it helps to have a way to calculate the total time it will take everyone to complete the job given their different rates of work. In this section, Dr. Math will look at some of these rate of work problems.

Ratio of Rates and Work

Dear Dr. Math,

Here's the question:

> Four skilled workers can do a job in 5 days. Five semiskilled workers can do the same job in 6 days. How long does it take 1 semiskilled and 2 skilled workers to do the job together?

I've tried this question several times using inverse proportion, but it doesn't work out.

—Carissa

Dear Carissa,

Let's look at each type of worker separately, with the goal of determining what fraction of a job one such worker will do in 1 day:

> 4 skilled workers can do a job in 5 days.

Since it takes them 5 days, they do $\frac{1}{5}$ of the job in each day. Since there are 4 of them, each does $\frac{1}{20}$ of the job in 1 day.

> 5 semiskilled workers can do the same job in 6 days.

Since it takes them 6 days, they do $\frac{1}{6}$ of the job in each day. Since there are 5 of them, each does $\frac{1}{30}$ of the job in each day.

So, the rate of work for a skilled worker is $\frac{1}{20}$ of the job per day, and the rate of work for a semiskilled worker is $\frac{1}{30}$ of the job per day. Since we have 1 skilled worker and 2 semiskilled workers, the combined rate of work is

$$\frac{1}{20}+\frac{1}{30}+\frac{1}{30}=\frac{3+2+2}{60}=\frac{7}{60} \text{ of the job per day}$$

Since rate of work is defined as the quotient of work divided by time:

$$r = \frac{w}{t}$$

time must be the quotient of work divided by rate:

$$t = \frac{w}{r}$$

When we divide one job by $\frac{7}{60}$ of the job per day, the answer is $\frac{60}{7}$ of a day, or $8\frac{4}{7}$ days.

Checking our answer,

In $\frac{60}{7}$ days, the skilled worker will do $\frac{60}{7} \cdot \frac{1}{20}$, or $\frac{3}{7}$, of the job.

In $\frac{60}{7}$ days, each semiskilled worker will do $\frac{60}{7} \cdot \frac{1}{30}$, or $\frac{2}{7}$, of the job. So two semiskilled workers will do $\frac{4}{7}$ of the job.

Together, the whole team will do $\frac{7}{7}$ of the job in $\frac{60}{7}$ days.

—Dr. Math, The Math Forum

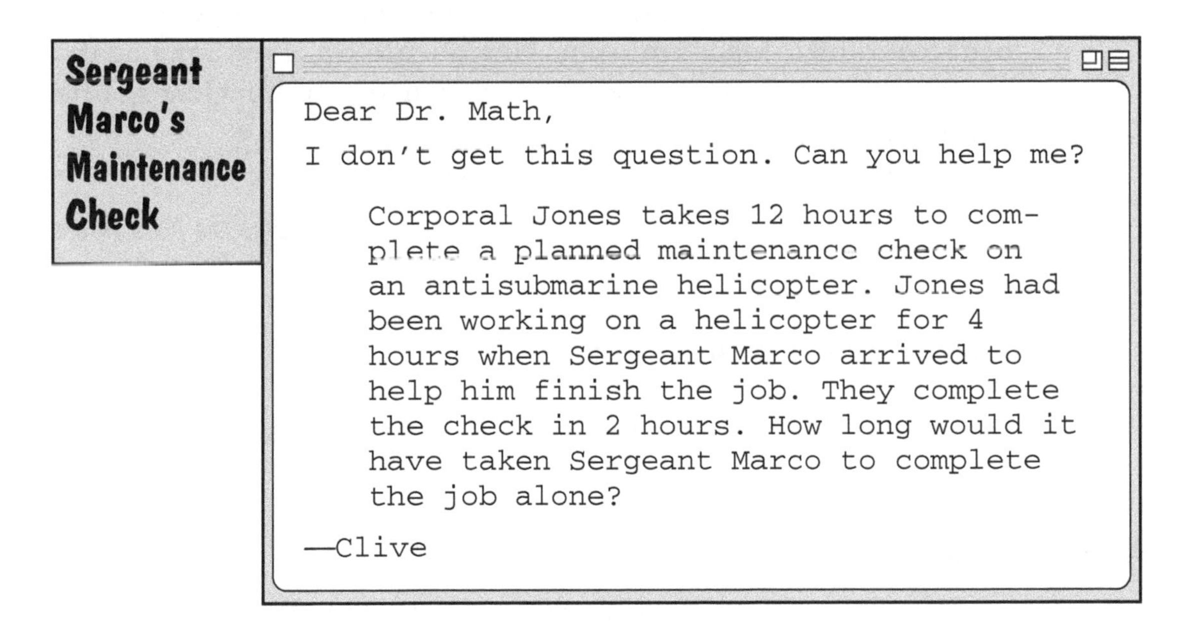

Sergeant Marco's Maintenance Check

Dear Dr. Math,

I don't get this question. Can you help me?

> Corporal Jones takes 12 hours to complete a planned maintenance check on an antisubmarine helicopter. Jones had been working on a helicopter for 4 hours when Sergeant Marco arrived to help him finish the job. They complete the check in 2 hours. How long would it have taken Sergeant Marco to complete the job alone?

—Clive

Hi, Clive,

This kind of problem is always tricky to think about. The trick is to make it into a rate problem, like distance = rate · time. But the Sergeant doesn't run a certain distance in an hour. He does a certain amount of work in an hour.

If Corporal Jones takes 12 hours to check a helicopter, then he does $\frac{1}{12}$ of the job in 1 hour. That's his rate: $\frac{1}{12}$ helicopter/hour.

In 4 hours, how much of the job has he finished? Just as distance = rate of speed · time, work = rate of work · time worked.

$$\frac{1}{12}\frac{\text{helicopter}}{\text{hour}} \cdot 4 \text{ hours} = \frac{1}{3} \text{ helicopter}$$

So, he has $\frac{2}{3}$ of the job left to do.

When Sergeant Marco gets to work, you have to assume that he works separately from Corporal Jones—they don't get in each other's way, and they don't help each other. (Sometimes just a little help can make a job go a lot faster—I hope I'm doing that for you!)

If you make this assumption, then Jones and Marco together work at a rate equal to Jones's rate plus Marco's rate. So, if you can find the rate at which they worked together, you can just subtract Jones's rate to find Marco's rate.

See if you can solve the problem now.

—Dr. Math, The Math Forum

Resources on the Web

Learn more about applications of numbers at these Math Forum sites:

Area and Perimeter

mathforum.org/alejandre/frisbie/dr.math.area.html

Links to letters from the Ask Dr. Math archives providing information on area and perimeter. Students are encouraged to follow the Cornell note-taking method.

Buying Tile

mathforum.org/alejandre/frisbie/buying.tile.html

Cooperative group activity designed to reinforce the concept of area. Two examples of student work are available.

ESCOT Problem: Marabyn

mathforum.org/escotpow/print_puzzler.ehtml?puzzle=44

mathforum.org/escot/marabyn.html

Students change the distance Marabyn rides the bus and the distance she walks to fit constraints about the distance and time that she walks.

Problems of the Week: Elementary School: Ratio and Proportion

mathforum.org /library/problems/sets/elem_ratio-proportion.html

Students learn to understand and use ratios and proportions to represent quantitative relationships. Students develop, analyze, and explain methods such as scaling and finding equivalent ratios to solve problems involving proportions.

Problems of the Week: Middle School: Distance, Rate, and Time

mathforum.org/library/problems/sets/middle_distance-rate-time.html

Distance-rate-time problems can be solved using simple variables, which offer a way to practice algebraic reasoning skills.

Problems of the Week: Middle School: Ratio and Proportion

mathforum.org /library/problems/sets/middle_ratio-proportion.html

Students learn to understand and use ratios and proportions to represent quantitative relationships. Students develop, analyze, and explain methods such as scaling and finding equivalent ratios to solve problems involving proportions.

What Is Area?

mathforum.org/alejandre/frisbie/one.inch.tiles.html

Collaborative group activity comparing area and perimeter using Hands-On Math software by Ventura Educational Systems, but the ideas could be adapted for use with other software or a Java applet.

Glossary

absolute value The absolute value of a number is its distance from zero on a number line.

algebra A branch of mathematics in which variables represent unknown quantities and can be manipulated in equations instead of and along with numbers.

approximate To approximate is to come close to an exact value—for example, $\frac{22}{7}$ approximates π, and 3.14159 is closer.

arithmetical operation One of the basic mathematical processes such as addition, subtraction, multiplication, or division.

axis A line that we use to find or draw points and shapes. If we want to talk about more than one axis, we say **axes**.

circumference The distance around the edge of a circle.

complex number A number made up of a real number plus an imaginary number, for example, $3 + 5i$. The real numbers are contained in the complex numbers, because real numbers are complex numbers whose complex parts are zero.

composite number A natural number with proper factors.

constant A quantity that doesn't change. Pi is a constant; so is 3 in the equation $y = \frac{4}{3}x + 3$.

coordinate A number that specifies (or helps specify) a location.

counting number Any number you'd use to count physical objects: 1, 2, 3, . . .

denominator The denominator of a fraction is the number on the bottom or below the fraction bar.

diameter The distance across the widest part of a circle; twice the radius.

distance Rate (a fractional quantity with distance in the numerator and time in the denominator, such as miles per hour) multiplied by time (hours, minutes, seconds, etc.).

distributive property of multiplication over addition For any real numbers a, b, and c: $a(b + c) = ab + ac$.

e The base of the natural logarithm, also known as Euler's number (because Euler named it, not because his last name begins with e!). You'll rely heavily on e when you take calculus.

equation A statement (which may be true or false) that two expressions or quantities are interchangeable.

equivalent Two things are equivalent if either can take the place of the other—for example, $\frac{1}{2}$ and $\frac{5}{10}$ are equivalent fractions. Equivalent expressions have the same value; equivalent equations have the same meaning.

estimate To estimate is to make a good guess at something, or to get a rough but not exact answer. If I add 3.17, 4.29, and 5.63, I get 13.09. I can estimate the sum as $3 + 4 + 6 = 13$.

evaluate To reduce an expression to a single numeric value by assigning values to its variables as necessary. For example, the expression $2(x + y)$ evaluates to 14 when the values of x and y are 3 and 4.

exponent A symbol placed to the right and above a mathematical term; a shorthand way of expressing multiple multiplications of a single quantity, as in $b \cdot b \cdot b \cdot b = b^4$.

expression A sequence of operations on a combination of numbers and/or variables, representing some quantity.

factor For natural numbers a, b, and c, if $a \cdot b = c$, then a and b are factors of c.

golden ratio $\frac{1+\sqrt{5}}{2}$, or approximately 1.618. . . . In a golden rectangle, the ratio of the long side to the short side is equal to the ratio of the sum of the long side and the short side to the length of the long side, which is equal to the golden ratio.

graph To graph is to locate and mark one or more points, lines, or curves on coordinate axes.

greatest common factor (GCF) The greatest or largest factor that two or more numbers have in common.

integer A whole number or a negative whole number.

inverse operation The inverse of something is that thing turned inside out or upside-down; the inverse of an operation undoes the operation—for example, division undoes multiplication.

irrational number A number that cannot be written as the *ratio* of two integers. The decimal representation of an irrational number (such as pi, *e*, or the square root of 2) neither terminates nor repeats.

least common denominator (LCD) The smallest number for which two or more given denominators are divisors. For example, 6 is the LCD for the fractions $\frac{1}{2}$ and $\frac{1}{3}$.

least common multiple (LCM) The least or smallest number, other than zero, that is a multiple of two or more given numbers.

linear equation An equation with two variables that describes a line. Technically, it's called an equation of degree 1, because the variables each have exponents of 1 (which don't usually need to be written out—that is, x^1 is the same as x).

natural number Another name for **counting number.**

negative number A number less than zero. We use the minus symbol for negative numbers, for example, –10 means negative ten. You can imagine this as a temperature 10 degrees below zero.

numerator The numerator of a fraction is the number on the top, or above the fraction bar.

order of operations The convention that operations be performed in a specific order within expressions:

1. Parentheses
2. Exponents
3. Multiplication and Division, left to right
4. Addition and Subtraction, left to right

ordered pair A list of two numbers where the order of the numbers is important. We write the coordinates of a point as an ordered pair inside parentheses. Some examples of ordered pairs are (1, 2), (–4, –5), (2, –7), and (–6, 3). Note that (1, 2) and (2, 1) are *different* ordered pairs.

pi Pi (π) is a constant, defined as the ratio of a circle's circumference to its diameter.

positive A number greater than zero. Sometimes we use the plus symbol for positive numbers, for example, +10 or 10 means positive ten.

prime factor A factor that is also a prime number.

prime number A prime number is a positive number whose only positive integer factors are 1 and itself.

proper factors The factors of a natural number, other than itself.

proportion A statement that two ratios are equal. They're sometimes read as "*a* is to *b* as *x* is to *y*."

quadrant The two axes divide the plane into four quarters, called quadrants. They are numbered from 1 to 4, starting in the upper right-hand corner and moving counterclockwise.

radius Half the diameter.

rate Distance (given in units such as miles, feet, kilometers, meters, etc.) divided by time (hours, minutes, seconds, etc.).

ratio A pair of numbers used to make a comparison, often written as a fraction—for example, $\frac{2}{3}$—but can be written as 2:3 or 2 to 3.

rational number Any number that can be expressed as a fraction, or *ratio,* of one integer to another.

real number Any rational or irrational number.

repeating decimal A number whose decimal part contains a digit or group of digits that repeat forever; 0.333 . . . is a repeating decimal, as is 3.269269269. . . . The repeated digit or group of digits may be written with a bar over the repeated numbers. For example, $3.\overline{269}$ is the same number as 3.269269269. . . . Note that 0.123456789101112131415 . . . is not a repeating decimal.

scientific notation The expression of a number as a number from 1 to 10 (not including 10) multiplied by a power of 10. For example, the scientific notation for 3,476 is $3.476 \cdot 10^3$.

solve To reduce an equation to the form "variable = expression or quantity," where the expression or quantity does not contain the variable. For example, the equation $2x + 4y - 6 = 0$ when solved for x is $x = 3 - 2y$. When solved for y it is $y = \frac{(3-x)}{2}$.

square root The square root of a number x is the number such that, when squared, it gives x. For example, 2 is the square root of 4.

symbolic form To put something into symbolic form is to translate it into numbers and symbols, as in an equation.

time Distance (given in units such as miles, feet, kilometers, meters, etc.) divided by rate (a fractional quantity with distance in the numerator and time in the denominator, such as miles per hour).

variable A symbol that stands for an unknown quantity in a mathematical expression or equation.

Index

D

E

F

G

H

I

J

Q

R

S

T

U

V

W

Y

Z

CPSIA information can be obtained
at www.ICGtesting.com
Printed in the USA
BVHW02s0454291217
503809BV00008B/16/P